PROJECT AIR FORCE

T0210378

F-35 Block Buy—An Assessment of Potential Savings

Appendix B, Historical Case Studies of Multiyear Procurement and Block Buy Contracts

Mark A. Lorell, Abby Doll, Thomas Whitmore, James D. Powers, Guy Weichenberg

Prepared for the United States Air Force

For more information on this publication, visit www.rand.org/t/RR2063z1

Library of Congress Cataloging-in-Publication Data is available for this publication.
ISBN: 978-0-8330-9836-8

Published by the RAND Corporation, Santa Monica, Calif.

© Copyright 2018 RAND Corporation

RAND® is a registered trademark.

www.rand.org

Preface

The F-35 Lightning II is the most expensive acquisition program in the U.S. Department of Defense. It is intended to replace several fighter and attack aircraft for the U.S. Air Force, Navy, and Marine Corps, as well as those from a number of partner allied nations. The U.S. military services and partner nations are keenly interested in ways to reduce the cost of the program. The F-35 Joint Program Office asked RAND Project AIR FORCE (PAF) to analyze what savings might accrue to the program if three upcoming lots of aircraft were to be procured under a single block buy (BB) contract as opposed to multiple annual contracts. Similar to multiyear procurement contracting, BB contracting should provide the prime contractors and their suppliers the incentive and ability to leverage quantity and schedule certainty and economies of scale to generate savings that would not be available under annual single-lot contracting.

This online appendix presents a set of case studies that PAF conducted to understand how historical weapon system programs have utilized multiyear procurement and BB contracts. The results inform key aspects of the main analysis and provide a broader context in which to understand how these contracting approaches have been exercised in the past.

As with the main report, the research reported here was sponsored by Lt Gen Christopher Bogdan, Program Executive Officer for the F-35 Lightning II Joint Program Office, and was conducted within the Resource Management Program of PAF. This document should be of relevance to those involved in the F-35 program and to those interested in methodologies for assessing cost savings in BB and multiyear procurement contracts.

RAND Project AIR FORCE

RAND Project AIR FORCE (PAF), a division of the RAND Corporation, is the U.S. Air Force's federally funded research and development center for studies and analyses. PAF provides the Air Force with independent analyses of policy alternatives affecting the development, employment, combat readiness, and support of current and future air, space, and cyber forces. Research is conducted in four programs: Force Modernization and Employment; Manpower, Personnel, and Training; Resource Management; and Strategy and Doctrine. The research reported here was prepared under contract FA7014-16-D-1000.

Additional information about PAF is available on our website: www.rand.org/paf/

This report documents work originally shared with the U.S. Air Force on July 1, 2016. The draft report, issued on September 30, 2016, was reviewed by formal peer reviewers and U.S. Air Force subject-matter experts.

Contents

Figures and Tables

Figures

Tables

Summary

This appendix discusses how historical multiyear procurement (MYP) and block buy (BB) contracts have been implemented and how they compare with each other. As context for RAND Project AIR FORCE's (PAF's) analysis of potential savings in an F-35 BB contract, we examined 28 historical multiyear contracts,[1] spanning 17 different weapon systems (15 aircraft and two naval vessels). This appendix outlines our methodology and data sources for analyzing historical MYP and BB contracts, provides a high-level overview of trends we observe across the case studies, and then provides in-depth discussion of the more-recent historical multiyear program case studies.

Each case study features a brief program summary, a description of the program's use of MYP or BB contracts, a comparison between the program's use of these contracts and the proposed F-35 BB contract, and observations that can be applied to the F-35 program. Looking across the case studies, we observe three main trends:

- There are significant differences among the key elements characterizing historical multiyear aircraft programs (e.g., program length, program size, annual production numbers, amount of economic order quantity [EOQ] funding or cost reduction initiative [CRI] funding). Each program must be evaluated in depth and on its own terms with respect to its unique characteristics. Notwithstanding this great diversity, there is a consensus among most program officials and subject matter experts that EOQ and CRIs are important drivers of savings on most MYPs.

- Comprehensive and in-depth analysis of contractor cost structure, particularly on the lower tiers, beyond what has typically been done in the past, can substantially increase estimated program savings, particularly during the contract negotiation phase.

- Development of an appropriate baseline estimate of the likely cost of comparable single-year contracts is difficult but crucial, as it is the basis for determining government estimated multiyear savings.

The remainder of this appendix details the case studies and lessons specific to each.

---

[1] We use the term *multiyear contract* as a generic term covering both formal MYP and BB contracts. Appendix A in RR-2063-AF discusses the history of multiyear contracting and details the differences between BB and MYP contracts.

Abbreviations

A/C	avionics/cockpit
AP	advanced procurement
BB	block buy
CAPE	Cost Assessment and Program Evaluation
CAPEX	capital investment incentives
CPIF	cost plus incentive fee
CPFF	cost plus fixed fee
CRI	cost reduction initiative
CV	carrier variant
DCMA	Defense Contract Management Agency
DoD	U.S. Department of Defense
EB	Electric Boat
EOQ	economic order quantity
EPA	economic price adjustment
EVM	Earned Value Management
FAR	Federal Acquisition Regulations
FFP	firm fixed-price
FMS	Foreign Military Sales
FP-EPA	fixed price with economic price adjustment
FPIF	fixed-price incentive fee (firm target)
FRP	full-rate production
FY	fiscal year
GAO	U.S. General Accounting Office (later Government Accountability Office)
LCS	Littoral Combat Ship
LMA	Lockheed Martin Aeronautics
LRIP	low-rate initial production

NDAA	National Defense Authorization Act
MDAP	major defense acquisition program
MM	multimission
MYP	multiyear procurement
OSD	Office of the Secretary of Defense
P&W	Pratt & Whitney
PB	President's Budget
RFP	request for proposals
RR	Rolls-Royce
SAR	Selected Acquisition Report
SSN	nuclear submarine
TINA	Truth in Negotiations Act
TY	then-year
UCA	Undefinitized Contract Action
UK	United Kingdom

Appendix B: Historical Multiyear Procurement Programs: Overview and Case Studies

This appendix discusses how historical multiyear procurement (MYP) and block buy (BB) contracts have been implemented and how they compare with each other. As context for our analysis of potential savings in an F-35 BB contract, we examined 28 historical multiyear contracts, including 17 different weapon systems (15 aircraft and two naval vessels). This appendix outlines our methodology and sources for analyzing historical MYP and BB programs, provides a high-level overview of trends we observe across the case studies, and then provides in-depth discussion of the more recent historical multiyear program case studies.

Methodology

Programs Included in the Analysis

Table B.1 lists the programs included in the historical analysis. In total, we reviewed 17 programs, encompassing 26 separate formal MYP contracts and two BB contracting efforts. These include all of the fixed-wing aircraft and most rotary-wing aircraft major defense acquisition programs (MDAPs) since the late 1970s that have used formal MYP contracts.[2] The fixed-wing aircraft programs included four fighter/attack aircraft (F-16, AV-8B, F/A-18E/F/G, F-22), one bomber (B-1B), three tactical transport/cargo aircraft (C-2, C-17, C-130J), two aerial warning and control aircraft (E-2C and E-2D), and one aerial tanker (KCH-10). The remaining three aircraft were multimission (MM) rotary-wing aircraft. Two Navy ship programs—the nuclear submarine (SSN) 774 *Virginia*-class attack submarine and the Littoral Combat Ship (LCS) surface combatant—were included because they are the only prior BB programs ever to have been implemented.[3] The U.S. Air Force was the lead service for six of these programs and involved in a seventh (V-22). The Navy was the lead service on all the remaining programs except one (CH-47F), which was managed by the U.S. Army. The Army also was a critical player on the MH-60R/S program, which was run cooperatively with the much larger Army UH-60M program.[4]

---

[2] The only aircraft MDAP MYP since 1990 that is not reviewed here is the Army AH-64 Apache attack helicopter MYP. For a comprehensive list of all MYP contracts approved by Congress between 1990 and 2015, see Ronald O'Rourke and Moshe Schwartz, *Multiyear Procurement (MYP) and BB Contracting in Defense Acquisition: Background and Issues for Congress*, Washington, D.C.: Congressional Research Service, R-41909, March 4, 2015.

[3] The *Virginia*-class submarine program also had multiple formal MYP contracts that we did not examine in detail.

[4] The Navy H-60 variants are procured jointly with the Army H-60 variants under a common contract. The Army leads the negotiation for a single MYP contract, which applies to both military services. However, the Navy calculates and reports its MYP budget justification package with estimated savings and funding separately from the

Table B.1. MYP and BB Contracts Reviewed by RAND

Program	Lead Service	System Type	MYP or BB	Number of MYPs/BBs
F-16	Air Force	Fighter	MYP	3
F/A-18E/F/G	Navy	Fighter	MYP	3
F-22	Air Force	Fighter	MYP	1
AV-8B	Navy	Fighter (VSTOL)	MYP	1
B-1B	Air Force	Bomber	MYP	1
E-2C	Navy	AWACS	MYP	2
E-2D	Navy	AWACS	MYP	1
C-2	Navy	Cargo	MYP	1
C-17	Air Force	Cargo	MYP	2
C-130J	Air Force	Cargo	MYP	2
KC-10	Air Force	Aerial refueling	MYP	1
V-22	Navy	Tilt rotor MM	MYP	2
CH-47F	Army	Rotor cargo	MYP	2
MH-60R/S	Navy	Rotor MM	MYP	2
MH-60R/S avionics/cockpit (A/C)	Navy	A/C	MYP	2
Virginia-class SSN	Navy	Attack submarine	BB	1
LCS	Navy	Surface combatant	BB	1

NOTES: The Navy MH-60 variants are procured jointly with the Army H-60 variants under a common contract. The Army leads the negotiation for a single MYP contract that applies to both military services. However, the Navy calculates and reports its MYP savings and funding, as well as its President's Budget (PB) exhibit data, separately from the Army, and sometimes reports it separately. The Navy variants have their own SARs. Therefore, we have narrowed our examination of the program to the Navy variants to reduce complexity, while keeping in mind the benefits (and potential pitfalls) of a joint production program with the Army. AWACS = Aerial Warning and Control System; VSTOL = vertical short takeoff and landing.

Data Sources

To provide a common baseline of format and approach for comparing estimated program costs and savings, we sought to use the standardized MYP PB exhibits or service justification packages developed by the U.S. Department of Defense (DoD) and the lead service, which must be formally presented to Congress to receive congressional authorization to implement an MYP. The final negotiated contract and savings estimate can differ significantly from the estimate presented in PB exhibits or justification packages, but we were unable to find easily available official sources for final

---

Army. The Navy variants have their own Selected Acquisition Reports (SARs). The two MH-60R/S MYP examined were actually part of the larger Army H-60 multiyear series; they constituted part of the MYP VII and MYP VIII of the H-60 multiyear series. However, the two Navy MH-60 MYPs were the only two that included both the MH-60R and MH-60S variants, and that were accompanied by separate all-Navy MYP contracts for the avionics and cockpit equipment. Therefore, we have narrowed our examination of the program to the Navy variants to reduce complexity, while keeping in mind the benefits (and potential pitfalls) of participation in a much larger joint MYP production program with the Army.

negotiated contract values that are directly comparable in content or format to the PB exhibit, or across programs.[5] Therefore, for purposes of direct comparison among programs, we used the formal PB exhibits or justification packages for each program whenever possible. For a rough comparison to actual program outcomes, we used the official data presented in SARs, which are formally reported to Congress each year as required by statute.

We chose to make an analytical distinction between older MYPs from the 1970s and 1980s and newer MYPs and BBs from the past 20 years.[6] For the older programs, it is much more difficult to obtain data from the program offices or even MYP PB exhibits. Reliable, publicly available information can be sparse, particularly for such smaller programs as the AV-8B, B-1B, C-2, and KC-10. In addition, these programs took place in an earlier acquisition regulatory environment (as well as an earlier technological environment), and thus might not be as relevant to the F-35 BB analysis as more recent programs. Thus we deemphasized these programs and focused on the more recent ones.[7] For some of these programs, we were able to draw on insights from prior RAND research projects, including RAND's 2007 assessment of the F-22 MYP.[8]

We also conducted extensive qualitative case study analyses of the majority of the programs listed in Table B.1. Our information came from numerous sources, summarized in Table B.2. In many cases, we visited the program office and interviewed senior cost, contracting, and other acquisition officials. We also used information gained from interviews conducted with senior contractor officials.[9] We obtained official program documentation from many of the program offices or from other DoD sources. We also reviewed relevant published articles on the programs from the most reputable industry publications.

Analysis Methods

We used two quantitative methods to analyze the historical programs. First, we conducted a regression analysis on a selected group of key program characteristics, which (based on expert

---

[5] For example, the SARs do not report contract costs in the same format or with parallel content as the PB exhibits. Price Negotiation Memoranda document the contract cost negotiations in great detail. These documents are produced by the program office contracting officials. However, these documents contain substantial proprietary and other sensitive data, and are difficult to obtain. Given our resource constraints, we were able to acquire the Price Negotiation Memoranda for only one program.

[6] We made an exception for the F-16 for two reasons. First, because of the data-gathering efforts of past research efforts, we possess detailed information on the budget exhibits as well as other estimates and information for the three F-16 MYPs. Second, with only two other historical fighter MYPs to examine, we thought it was important to include the F-16 in the analysis when comparing to the F-35.

[7] Again, we make an exception for the F-16, for which we have considerable data, and which is only one of three fighters to experience an MYP.

[8] See Obaid Younossi, Mark V. Arena, Kevin Brancato, John C. Graser, Benjamin W. Goldsmith, Mark A. Lorell, Fred Timson, and Jerry M. Sollinger, *F-22A Multiyear Procurement Program: An Assessment of Cost Savings*, Santa Monica, Calif.: RAND Corporation, MG-664-OSD, 2007.

[9] Most interviews with program office and contractor officials were conducted in person, but a small number involved teleconferences.

opinion) we reasoned might be linked to the level of savings estimated. This approach is explained in more detail later. For some programs, we also compared estimated costs just before the MYP with actual MYP costs to determine whether the MYP achieved its projected cost savings goals.

Table B.2. RAND Sources of Information for MYP/BB Program Case Studies

Program	U.S. Government/ Program Office Documentation	Contractor Interviews, Prior Research	Program Office Interviews, Prior Research	Program Office Interviews, Current Research
F-16	Yes	Yes	Yes	No
F/A-18E/F/G	Yes	Yes	Yes	No
F-22	Yes	Yes	Yes	Yes
AV-8B	Yes	Yes	No	No
B-1B	Yes	No	No	No
E-2C	Yes	Yes	Yes	No
E-2D	Yes	Yes	Yes	No
C-2	Yes	No	No	No
C-17	Yes	Yes	Yes	Yes
C-130J	Yes	Yes	Yes	Yes
KCH-10	Yes	No	No	No
V-22	Yes	Yes	Yes	Yes
CH-47F	Yes	Yes	Yes	Yes
MH-60R/S	Yes	No	No	No
MH-60R/S A/C	Yes	No	No	No
Virginia-class SSN	Yes	Yes	Yes	Yes
LCS	Yes	Yes	Yes	Yes

Overview of Insights Across Case Studies

Relating Program Characteristics and Cost Savings

We first sought to determine whether there was a correlation between program characteristics (e.g., program length, program size, annual production numbers, amount of economic order quantity [EOQ] funding or cost reduction initiative [CRI] funding) and the level of estimated savings for historical MYP and BB programs. Table B.3 summarizes some of the key attributes we considered. It includes all of the fixed-wing MYP contracts over the last two decades, as well as all the rotary-wing MYP contracts, with the exception of the AH-1D Cobra attack helicopter. In addition, we have included the three F-16 MYP contracts, because the F-16 is one of only three jet fighters that have used the MYP contracting approach in the past. The table also includes parallel characteristics for the proposed F-35 BB contract, highlighted in the last row.

Table B.3. Overview of Selected Characteristics of Multiyear Aircraft Programs

Program (Year Commenced)	Estimated Savings as % of Prime Contract	Overall Length of MYP (Years)	Number of Prior Production Lots (LRIP & FRP)	Total Number, Annual/Production Rate	Estimated Prime Contract Value (TY $B)	EOQ Funding as % of Contract	Government CRI Funding as % of Contract	Contract Type
F-16 I 1982	7.7	4 1982–1985	3 LRIP 2 FRP	480 120	2.9	7	0	FPIF
F-16 II 1986	8.4	4 1986–1989	3 LRIP 6 FRP	720 180	3.9	2	0	FPIF
F-16 III 1990	5.7	4 1990–1993	3 LRIP 10 FRP	630 120–180	4.3	Unknown	Unknown	FFP
C-17 I[a] 1997	5.5	7 1997–2003	6 LRIP 2 FRP	80 8–15	14.2	2.1	2.5	FFP
E-2C I 1999	8.3	5 1999–2003	0 LRIP 0 FRP	21 3–5	1.3	32.2	0	FFP
F/A-18E/F I 2000	7.4	5 2000–2004	3 LRIP 5 FRP	222 36–42	8.8	0.96	1	FPIF
C-17 II 2003	10.8	5 2003–2007	6 LRIP 9 FRP	60 7–12	9.7	7	2.5	FFP EPA
C-130J I 2003	10.9	6 2003–2008	1 LRIP 7 FRP	62 1–30	4.2	3.3	0	FFP EPA
E-2C II 2004	7.2	4 2004–2007	0 LRIP 5 FRP	8 2	0.8	10.9	0	FFP
F/A-18E/F II 2005	10.9	5 2005–2009	3 LRIP 5 FRP	210 42–73	8.9	0	2	FFP
F-22 2007	4.5	3 2007–2009	9 LRIP 2 FRP	60 20	7.3	4	0	FFP EPA
MH-60R/S I 2007	4.7	5 2007–2011	3 LRIP 1&6 FRP	260 (144+116)	3.5	0	0	FFP
CH-47F I 2008	10	5 2008–2012	3 LRIP 3 FRP	215 35–47	4.2	0	<1	FFP
V-22 I 2008	4.1	5 2008–2012	9 LRIP 2 FRP	185 33–36	10.1	1.6	1.7	FPIF

Program (Year Commenced)	Estimated Savings as % of Prime Contract	Overall Length of MYP (Years)	Number of Prior Production Lots (LRIP & FRP)	Total Number, Annual/ Production Rate	Estimated Prime Contract Value (TY $B)	EOQ Funding as % of Contract	Government CRI Funding as % of Contract	Contract Type
F/A-18E/F III 2010	9.2	5 2010–2014	3 LRIP 10 FRP	174 40	7.6	0	1.3	FPIF
MH-60R/S II 2012	9.2	5 2012–2016	3 LRIP 6 & 11 FRP	193	3.4	0	0	FFP
CH-47F II 2013	10.0	5 2013–2017	3 LRIP 8 FRP	155 28–39	3.4	0	0	FFP
V-22 II 2013	11.6	5 2013–2017	9 LRIP 7 FRP	98 18–24	6.5	0.5	0	FPIF
C-130J II 2014	9.5	5 2014–2018	3 LRIP 19 FRP	79 8–28	5.8	3.9	Unknown	FPIF
F-35 BB 2018	6.5[b]	3 2018–2020	11 LRIP 0 FRP	431 139–149	30.0	4	0.8	FPIF

NOTES: Except where noted, all numbers are based on budget exhibits from MYP justification packages. The content of the MYP contracts varies considerably from contract to contract, thus complicating direct comparisons. In most cases, but not all, the engine is excluded from the contract. For example, the F-16 MYP I contract excludes the engine and many avionics systems and other components. The F/A-18E/F MYP I excludes the engine. The F-22 MYP had two separate MYPs, one for the air vehicle and one for the engine. LRIP = low-rate initial production; FRP = full-rate production; TY $B = billions of then-year dollars; FPIF = fixed-price incentive (firm target); FFP = firm fixed-price; EPA = economic price adjustment.

Parallel characteristics for the proposed F-35 BB contract are highlighted in the last row (shaded in gray).

[a] No PB exhibit was available. These numbers are based on a U.S. Government Accountability Office (GAO) description of the budget exhibit. (Note that the office changed its name in 2004 to U.S. General Accounting Office (GAO).) See GAO, *C-17 Aircraft: Comments on Air Force Request for Approval of Multiyear Procurement Authority,* GAO/T-NSIAD-96-137, March 1996.

[b] This savings estimate is for the air vehicle, including government-funded CRIs.

A quick review of the MYP characteristics shows the extensive variation and diversity among these programs. Savings estimates varied from 4.1 percent for the V-22 MYP I to 11.6 percent for the V-22 MYP II, with most savings estimates clustering around the 8–11 percent range.

There is also considerable variation in program characteristics. Program lengths vary in this sample from three years for the F-35 and F-22 to seven years for the C-17 MYP I. One would intuit that longer programs would have greater opportunities for savings because of economies of scale. A similar observation might apply to the size of planned production quantities, which vary from a mere eight for the E-2C MYP II to up to 630 for the F-16 MYP III. Annual production rates might also affect cost savings, with programs varying from two aircraft per year for the E-2C MYP II to 180 per year for the F-16 MYP II and MYP III. The level of EOQ funding and government-financed CRIs could also be expected to have a significant impact on savings; among the cases in Table B.3, EOQ funding varied from 0 percent on as many as five of the programs to as high as 2.5 percent of the estimated MYP contract value for the C-17 MYP I. Many historical MYPs had no government-funded CRIs, while others had as much as 2.5 percent of the estimated MYP contract value in CRI funding from the government.

We also considered the maturity of each program; that is, how complete the development process was, and how many LRIP and FRP lots had been already contracted at the commencement of the MYP or BB. The expectation was that the more mature the program (suggesting that the design and production items would be stable and well-tested), the higher the possible savings. Interestingly, by statute, MYPs must only be applied to programs with mature and stable designs, as discussed in Appendix A of the main report. Yet historically, many of the most complex MYP aircraft development programs launched their first MYP very early in the production process, even when the development program was experiencing, or recently had experienced, serious design, development, and developmental testing challenges. For example, the F-16, C-17, and F/A-18E/F launched their first MYP contracts after only a few LRIP and FRP lots. In the case of the C-17 and V-22, both programs had also experienced considerable challenges during the development stage, some of which had not been entirely resolved at the beginning of their first MYPs. If MYP contracts started without fully stable designs, it is possible that potential savings could be reduced.

A problem with making comparisons among these cases is that sometimes there are large differences regarding which aspects of the program and components of the weapon system are covered by the multiyear contract. For example, in the case of the F-16 MYP I, only the airframe and selected subsystems and components are covered. There was no MYP for the engine. In the case of the F/A-18E/F MYP I and the C-17 MYP I–II, there were totally separate MYPs for the airframe and the engine. The F-22 contract also had separate airframe and engine MYPs, but because of the unique situation surrounding this MYP, we used an overall estimate of savings for both airframe and engine.[10] In the case of the E-2C MYP II, the engine was included in the overall MYP, and we have no way of factoring out the contribution of the engine to overall savings. For the MH-60R/S program, a completely separate MYP existed for the aircraft avionics and cockpit instrumentation, which we report in Table B.3. Since engine MYPs have historically had smaller percentage savings than airframe MYPs, including or excluding the engine can make a difference.

In the case of the F-16 MYP I, we have two formal estimates of savings, based on different assumptions for the program. The Air Force developed a formal estimate of savings in March 1981, for a total MYP airframe savings of 10.5 percent. Later, the Air Force changed assumptions regarding inflation rates and some factors of the program contract. As a result, the official Air Force estimate was updated and submitted to Congress in October 1981 to reflect these changes, leading the savings estimate of 10.5 percent to be reduced to 7.7 percent, a significant difference.[11] Thus, changes in assumptions and what is included in the MYP can change the estimated savings percentage significantly. It is very difficult to know what was included and all the assumptions in every one of our historical programs and to determine how to normalize all the cases to ensure "apples to apples" comparisons.

Finally, many of these programs varied considerably in the type and complexity of the development program. For example, aircraft such as the V-22 and the F-16 were new designs of complex aircraft that had just gone through design and development of the entire weapon system. Other programs, such as the E-2C, the CH-47F, and the MH-60R/S, entailed relatively minor design and airframe changes, combined with the insertion of new avionics and other subsystems to basic platform designs that had been in the inventory for years or even decades. In many respects, these latter programs were extensive modification efforts, rather than full-scale development of new platform systems, and thus were generally less complex. It is not clear what effects these differences might have on potential savings.

Given the extensive variation in key program characteristics, we decided to conduct a regression analysis of all these factors and others to see which might be most important in driving savings estimates for an MYP. The specific factors included the size of the contract, the contract length, the scale of EOQ funding, the scale of CRI government funding, the date of the contract award, the contract type, the total planned production numbers, the annual production rate, and the maturity of the program as determined by number of LRIP and FRP lots before the beginning of the MYP. *This regression analysis showed no statistically significant correlation*

---

[10] There is a formal PB exhibit for F-22, but owing to time considerations, it was never completed with actual data. All the places for data are left blank. See U.S. Air Force, *FY 2007 Budget Estimates, Aircraft Procurement, Air Force*, Vol. 1, February 2006. Based on interviews with the F-22 program office, we used the RAND estimate of most likely program savings (see Younossi et al., 2007).

[11] See GAO, *An Assessment of the Air Force's F-16 Aircraft Multiyear Contract*, GAO/NSIAD-86-38, February 1986; also see F-16 System Program Office, *Validation of Multiyear Savings Associated with the Production of 720 F-16 Aircraft (FY86–FY89 Requirements)*, unpublished paper, September 1986.

between any program attribute we selected and the level of estimated cost savings preceding the MYP. This does not necessarily disprove that these factors are linked to the level of estimated savings and that some may be more important than others. The differences among program characteristics and analytical assumptions may have been too great and our overall sample size too small to achieve statistically significant results from the regression analysis.

Based on our interviews with program office and industry officials, it also became clear that there is a wide divergence of opinion regarding which policies are most important for achieving savings in MYPs. Some officials stressed the importance of EOQ funding, while others (especially industry representatives) emphasized the criticality of government-funded CRIs. Still other program office officials were very skeptical of government-funded EOQ and CRIs, arguing that if the program and contract are structured appropriately, the contractor will be incentivized to do what is necessary on its own to achieve maximum savings. With the data available to us, we are unable to resolve these issues. But as Table B.3 demonstrates, programs without either EOQ or CRI funding were estimated to achieve substantial savings that did not vary in a statistically significant way from programs that did have government funded EOQ and/or CRI funding.[12]

First MYPs Compared with Second MYPs for the Same Weapon System

In the course of examining these historical MYP and BB case studies, we noticed an interesting phenomenon: For a significant number of programs that had more than one MYP contract, the savings estimate used in the formal justification documentation for the later MYP contract was often higher than it had been for the first MYP contract. Initially, this seemed counterintuitive. We reasoned that earlier in these programs, there would be more opportunities to take advantage of "low-hanging fruit" and that the task of finding cost-saving measures and processes and other efficiencies would be relatively easier in the early phases of a program than in later phases. The analogy to this line of reasoning would be the typical behavior of production unit cost in "learning" or "cost improvement" curves. At the beginning of programs, learning curves are typically quite steep as the design stabilizes and experience is gained in manufacturing and assembling the weapon system. Later, learning curves tend to flatten out as all the most obvious areas for efficiencies and learning are exploited, and further reductions in unit cost become more difficult. Therefore, we were intrigued by the apparent tendency of second MYPs to have higher estimated cost savings than first MYPs for the same system.[13]

---

[12] There is, however, a widespread consensus in the literature and among many program officials that EOQ funding and CRIs are the main drivers of savings.

[13] Note, however, that this was not a statistically significant characteristic with respect to savings.

This phenomenon is illustrated in Figure B.1, which compares eight recent aircraft MYPs that have had two separate multiyear contracts, as well as the Virginia-class SSN 776.[14] This includes all the aircraft MYPs listed in Tables B.1 and B.2.[15] Of the nine weapon systems shown, only two (E-2C and C-130J) had precontract estimates that were lower in the second MYP than in the first.[16] Of these, the E-2C was an unusual MYP that had extremely small procurement numbers, particularly in MYP II (only eight aircraft total). The C-130J was unusual in that more than ten years passed between the start of the first MYP and the start of the second. Furthermore, during contract negotiations, the savings estimate for the second MYP contract was significantly increased. In the case of all other programs, the MYPs were sequential. Other special circumstances, discussed below, show that this second C-130J MYP was actually more similar to the others in having an increased savings estimate after negotiation of the final contract. The CH-47F was also a case where the cost savings estimate for the second MYP increased considerably during contract negotiations. The remaining six programs show increases in the PB exhibit or in service justification savings estimates for the second MYP prior to contract negotiations—sometimes substantial increases, such as in the case of the C-17, MH-60R/S, and V-22. In the following pages, we examine what explains these differences.

---

[14] We originally examined the SSN 776 because it is one of the only two BBs ever implemented, the other being the LCS. In the course of interviewing program officials, we learned about MYPs used after the BB. We gained some insights on why the savings increased in the second MYP, and therefore included the SSN 776 in our table of programs with at least two sequential MYPs. We discuss this issue further later on.

[15] The two MH-60R/S Avionics/Cockpit MYPs shown in Tables B.1 and B.2 are excluded because we have no reliable estimated cost savings data for the first MYP. We do have revealing information on the second MYP, however.

[16] One, the CH-47F, had two consecutive MYPs with the same savings estimate.

Figure B.1. Savings Estimates for First vs. Second MYPs for Selected Systems

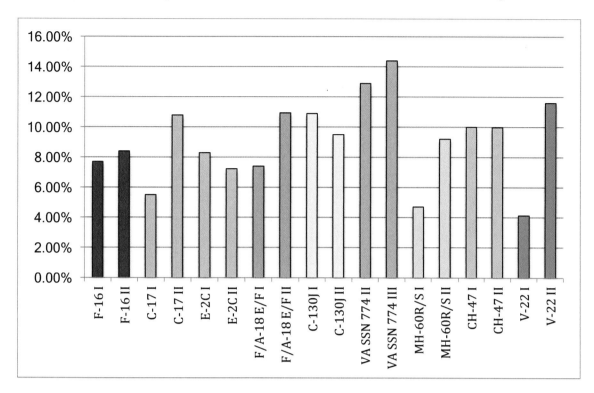

NOTE: Savings estimates are from PB MYP exhibits, military service MYP budget justification packages, or similar data before negotiation of the actual MYP contract.

Prenegotiation Savings Estimates vs. Negotiated Contract Settlements

A possible explanation for the above observation began to emerge when we were able to learn more details regarding the development of the second MYP savings estimates and acquired information on the savings estimated in the negotiated contract settlements. Unfortunately, we had access to this type of information on only a few programs, but the information is consistent across these programs. On two contracts, the MH-60R/S and the MH-60 R/S A/C MYP, which were fully separate multiyear contracts, a unique set of circumstances led to the formal publication of MYP justification packages before and after the negotiation of the actual contract; thus, we can directly compare the two. In the case of the MH-60R/S A/C program, we also can compare the progression of the precontract cost savings estimate over several years based on different published budget justification documents. We also have published data that enables a comparison of the C-130J MYP II prenegotiation savings estimate from the budget justification package, and the savings estimate based on the final negotiated contract, which also appears in a formal budget exhibit. In two other cases, we know the final negotiated savings estimates based on interviews with the program offices. In one of these cases, we also know the cost savings objective of the government and the contractor at the beginning of the negotiations.[17]

---

[17] We withhold the names of the programs at the program offices' request.

11

Documentation on all five of these programs shows significant increases in the cost savings estimate during final contract negotiations.

Figure B.2 shows this information for the MH-60R/S MYP II, the MH-60R/S A/C MYP II, and the C-130J MYP II. The information is from published budget documents, or, in the case of the C-130J negotiated savings percentage, from the SAR.[18] In all three cases where we have published estimates in similar formats, the negotiated contract savings estimate was greater than the formal budget MYP justification estimate—much greater in the case of two of the programs. The savings estimate for the negotiated contract for the MH-60R/S MYP II was two-and-a-half times greater than the prenegotiation official savings estimate; for the MH-60R/S A/C MYP II, the negotiated contract savings estimate was more than two times greater. In the case of the C-130J, the negotiated savings estimate rose from 9.5 percent in the prenegotiation estimate to 11.5 percent. Various assumptions changed between the prenegotiation estimate and the actual contract, but from what we know, none of these changes seems sufficient to account for the large change in the estimates, particularly for the first two programs. For example, in all three cases, the number of procurement units declined, but in the case of the MH-60 MYPs, these declines were small (from 193 to 182) and do not seem to be sufficient to explain the changes in the savings estimates. However, we do not know enough about these programs to determine with certainty what caused the large increase in the savings estimate during negotiations, and whether the causes among them were similar.

---

[18] DoD, *Selected Acquisition Report: C-130J SAR*, Washington, D.C., December 2015.

Figure B.2. Savings Estimates from Justification Exhibits Compared with Negotiated Contract Savings Estimates for Three Recent MYPs

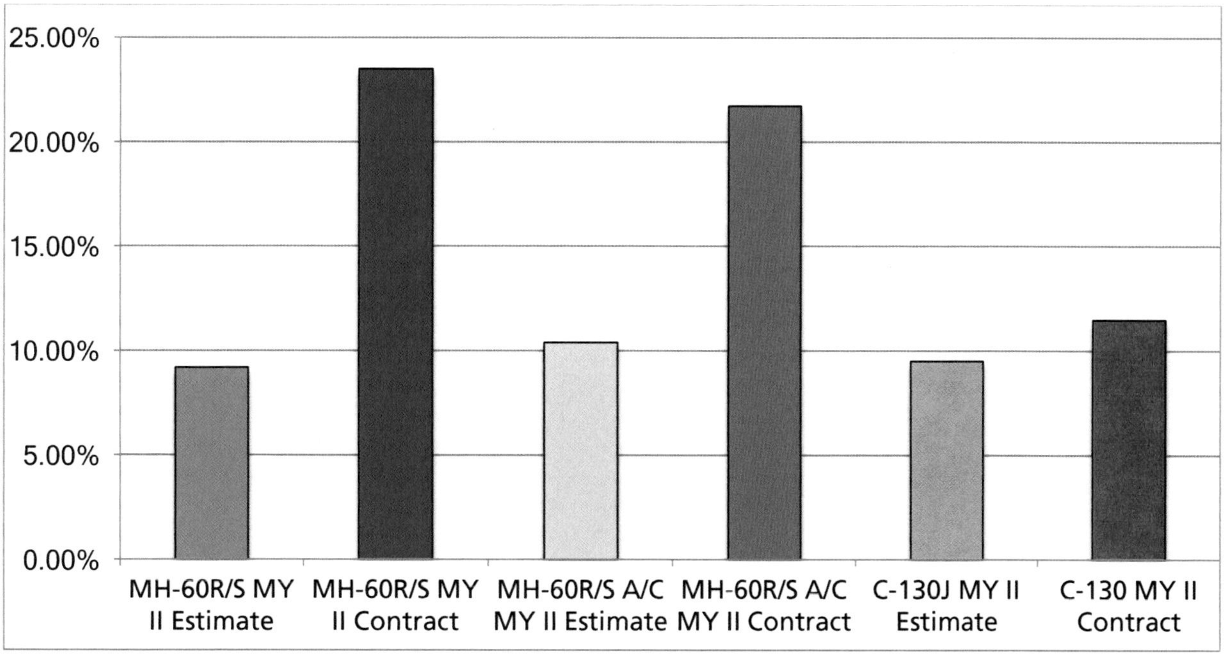

NOTE: In the case of the C-130J, the negotiated savings estimate is from the program SAR.

We do know more about two other recent MYP contracts, which were both follow-on MYPs that we were able to research in great depth, including extensive discussions with program officials. These two programs experienced outcomes very similar to the three programs already reviewed. Because of the sensitive nature of some of the information and details the program offices shared with us, we agreed not to identify the programs or the specific percentage savings. Nonetheless, the information we can share here is revealing. For one of the contracts—an MYP for program "X"—the negotiated savings estimate nearly doubled compared with the official prenegotiation savings estimate, as shown in Figure B.3, and we do have an extensive explanation why from the program office. This is also true of another MYP contract we examined in great detail, for program "Y." In this second program, we know the percentage savings estimate offered in the original prime contractor proposal, the savings objective for the program office going into the contract negotiations, and the actual savings percentage estimate for the final negotiated contract settlement. Note that the government went into the contract negotiations with a very aggressive position, as shown in Figure B.3, with a savings percentage goal nearly four times higher than that offered by the prime contractor in the original proposal. The final negotiated contract settlement resulted in a savings estimate about two and a half times higher than the original contractor proposal. All of these savings estimates were much higher than earlier budget justification package estimates for earlier MYPs for the same system.

Figure B.3. Savings Estimates and Negotiating Positions Before Negotiations Compared with Negotiated Contract Savings Estimates for Two Recent MYPs

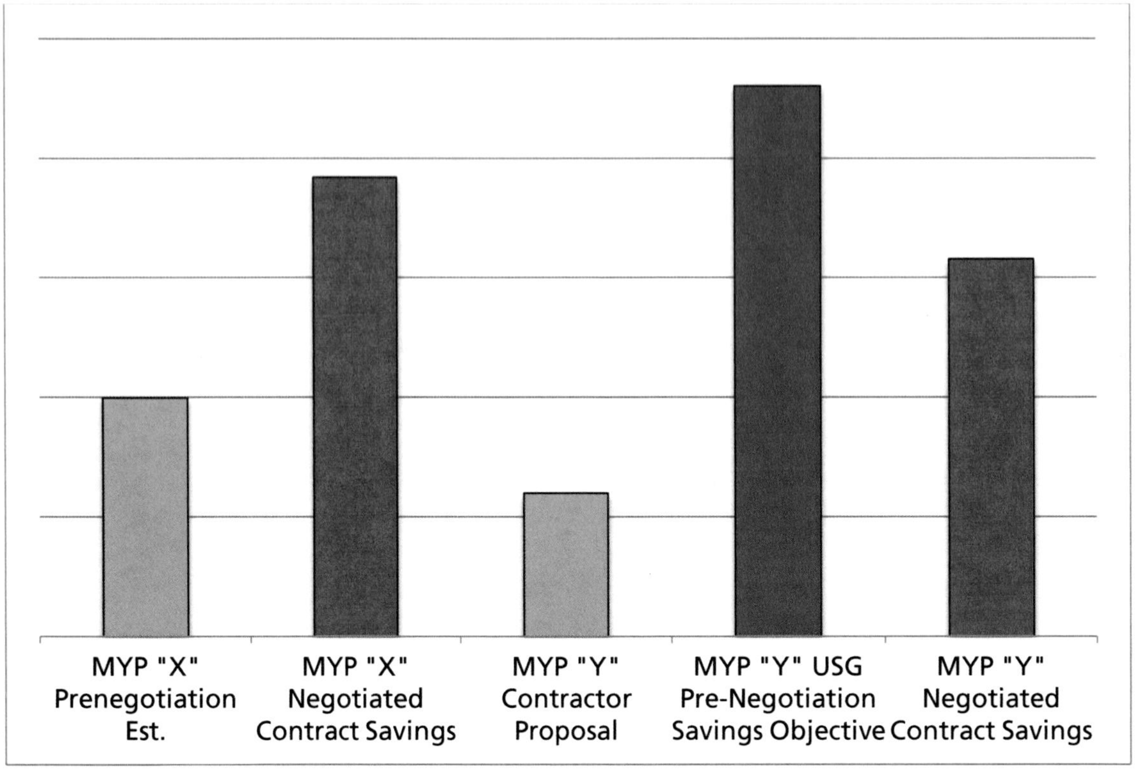

NOTE: USG = U.S. government.

While we have detailed quantitative and qualitative information on only these two programs, we believe what we learned may provide insight, or at least a hypothesis, explaining why second MYPs often produce savings estimates higher than first MYPs for the same system, and why negotiated settlements (for the five programs for which we have data) also resulted in higher savings estimates than the prenegotiation estimates.

The Critical Importance of Prenegotiation Fact-Finding

The increase in estimated savings between first and second MYPs and from the prenegotiation phase to the contract settlement appear to be linked to the government gaining greater knowledge and insight into actual contractor costs, especially at the supplier and lower-tier levels.[19] Our interviews with program officials representing the MYPs for programs "X" and "Y"—as well as with other MYP contract officials and experienced personnel at the Defense Contract Management Agency (DCMA)—strongly support this finding.

During the first MYP of at least three recent programs with multiple MYPs that we assessed in great detail, program officials observed that, based on the actual cost data received during

---

[19] Most MYP justification packages, as well as much of the published literature, identify EOQ purchasing at the supplier base, whether funded by the government or contractors, as a critical driver of MYP savings.

production through the standard DoD Cost and Software Data Reporting system, the prime contractor was consistently underrunning original estimated costs, and thus was making a higher profit than anticipated (because all multiyear contracts are fixed-price type contracts with the fixed price established during the contract negotiations). In several cases, before beginning negotiations for the second MYP, the program office and DCMA conducted a root-cause analysis to determine the origin and causes of the contractor cost underrun during the first MYP. Contracting officers concluded that the typical government approach to estimating savings before negotiations was flawed, permitting the contractor to reap savings higher than anticipated profits.

Historically, contracting officers typically look at original vendor quotes or bids to prime contractors (as well as overall prime bids) and compare the actual negotiated prime contractor prices and vendor prices (or "historical purchase orders") from prior contracts. During the postmortem reviews of the initial MYP for program "X," program officials concluded that historical purchase orders were inadequate, for two reasons. First, vendor bids can be significantly higher than the actual fixed-price contract negotiated later by the prime with vendors. As a result, prime contractors can negotiate with the government based on vendor bids, then go back and negotiate better deals with the vendors, thus realizing greater profits. Second, even if the government has obtained the actual negotiated vendor-prime contracts before negotiating the prime contract, the government may not have the actual comprehensive cost data for second- and third-tier suppliers—just prices rather than costs. Without knowing actual supplier costs, the government has difficulty estimating the amount of savings that is potentially available. As a result, the program office concluded that multiyear savings were retained by the contractors as extra profit, particularly on the lower tiers.

In short, DCMA concluded that, in many cases on MYPs, the government does not have adequate cost and pricing data when the prime contractor initially submits its proposal in response to the request for proposals (RFP) and the government begins contract negotiations with the prime contractor. The origin of this problem is differing interpretations of Part 15 of the Federal Acquisition Regulations (FAR). Prime contractors often argue that while subcontractor cost and pricing data must be provided to the government, it can be supplied at any time throughout the negotiations. DCMA concluded that the government needs the actual certified cost and pricing data that will apply to the actual program upfront when the proposal is submitted, or even before. Prime contractors often maintain that they can offer historical data in the proposal and then present initial quotes, and not actually negotiate the final contract with the supplier until after the government negotiations with the prime are concluded. Contractors often insist that they only need to certify the data they have on hand at the time of proposal submission to technically meet the FAR Part 15 legal requirement for certified cost and pricing data. The result is that the program office does not have adequate time to assess the final data before negotiations, which confers a large advantage to the prime contractor. After the fixed-price contract with the government is concluded, the prime contractor can then negotiate much tougher deals with suppliers, which raises the prime contractor's profits at the expense of savings theoretically obtainable by the government on MYPs. DCMA and program officials on at least three of the programs we examined in detail concluded that this is exactly what happened in the initial MYP for their weapon systems.

In all three of these cases, the government realized it had to be much better informed and spend much more time on fact-finding and obtaining actual production cost data, particularly from the lower-tier suppliers, before negotiations with the prime contractor. The problem is that this approach requires much more time, effort, and resources than are typically available in negotiating an annual contract.

Two of the program offices we interviewed in depth developed different approaches to this issue. The program "X" office, working closely with DCMA and other agencies, developed new instructions for the RFP to the prime contractor. The main goal was to receive high-quality and highly detailed certified cost and pricing data from the Truth in Negotiations Act (TINA)–compliant subcontractors (sole-source, noncompetitive proposals over $700,000) before the prime contractor proposal was submitted.[20] This affected a large number of suppliers, on the order of 80 or more. While the prime contractor balked and opposed these new instructions, the government exercised leverage by threatening Defense Contracts Audit Agency (DCAA) audits of the prime contractor business and cost systems for noncompliance. The program office permitted submission of detailed cost and pricing data from the subcontractors directly to the government rather than through the prime contractor, as is more common. The prime finally accepted this approach, although the negotiations for program "X" slipped by about two months as a result of contractor opposition, and also because of the extra time required for more-extensive fact-finding by the program office. According to the program office officials, however, the final negotiated savings estimate nearly doubled compared with the official prenegotiation savings estimate (as shown in Figure B.3) as a result of these measures.[21]

Program "Y" developed a slightly different but equally effective solution to this challenge. As with program "X," the program "Y" office came to the same conclusion that far more lower-tier contractor cost and pricing data are required before negotiations, and decided to do a very detailed assessment of 23 large vendors' cost structures. However, this required considerably greater personnel resources than were available to the program office. In response, the program office obtained outside military service assistance, gathered much more detailed labor rate information and other data from DCMA and DCAA, and, most important, hired a prominent consulting firm to help with the analysis. In addition, the program office negotiated with the prime contractor for independent assessment of key second-tier vendor bids. The goal was to

---

[20] Typically, lower-tier contractors submit data to the prime contractor, which then assesses those data and submits them with its proposal to the government.

[21] Information from DCMA interviews; DCMA, "Leveling the Playing Field, Fail-Safe RFP Proposal Instructions," undated briefing presented to RAND, Arlington, Va., March 20, 2015.

determine actual costs for both large second-tier vendors and their lower-tier suppliers. The assessments included detailed examination of labor rates, pensions costs, overhead, health costs, workload assumptions, and so forth. Based on this assessment of the lower tiers, which normally avoid direct government scrutiny, the program office felt it was in a much stronger position to negotiate with the prime contractor. This is because, with FFP contracts, many of the lower-tier vendors resist providing actual cost data to higher-tier vendors. As shown in Figure B.3, the prime contractor's cost savings before entering negotiations was modest. Backed by detailed knowledge of lower-tier costs, the government entered negotiations with a very aggressive savings estimate objective, more than three times greater than the contractor's offer. The final outcome at contract signing was a savings estimate about two and a half times higher than the prime contractor's opening position.

In the view of program officials, the government benefited significantly from the added research and work compared with the traditional approach of depending on lower-tier contractor bids to prime contractors or other historical cost data. It was the opinion of the program office financial and contracting officers that a rigorous approach like this—obtaining detailed cost information from the prime contractor and digging into price and cost realism at all lower tiers of suppliers—and in all categories (labor hours/rates, material, etc.) is key to obtaining significant savings on large contracts. And while this approach is not specific to a multiyear contract, its use for a multiyear contract is especially important because multiple lots are at stake and because a multiyear contract provides greater justification and time for committing the extra resources that such an approach requires. Finally, this approach could also be used for developing more-realistic savings estimates for the official justification package for Congress, although this would require more resources and time than needed for the traditional approach.

The Challenges of Evaluating Savings

So far, based on our review of the literature and the histories of 15 past aircraft MYP contracts, some with multiple MYP contracts, plus two ship programs with BB contracts, at least two important points have emerged: (1) the challenge of fairly comparing savings estimates from one program to another, because of the wide diversity of so many key program characteristics that contribute to savings; and (2) the critical importance of the contract negotiation phase, backed by a great depth of data and knowledge regarding costs on the lower tiers. These two elements make evaluating and comparing estimates across programs difficult. Even more challenging is assessing the actual savings achieved by past programs or likely to be achieved by future programs. In addition to the factors discussed above, this task is made even more difficult by the fact that all historical program savings claims are based on prenegotiation estimates or negotiated contract estimates, not on actual ex post facto cost results, and are compared with hypothetical estimates of comparable single-year contract baselines, not actual programs.

Convincingly demonstrating actual MYP cost savings would require pursuing two identical programs: the first using an MYP contract and the second using annual contracts. Even if we impose less restrictive conditions for achieving accurate estimates of actual savings, a nearly insurmountable problem is presented by the fact that DoD does not systematically collect and retain data on the development of the original prenegotiation savings estimate, nor does it do so for useful data on the actual achievement of those savings.[22] This task is further complicated by the inevitable program changes that take place during implementation of MYPs, such as changes in budget, quantities, rate, requirements, design, and modifications.[23]

Overview Summary

History and analysis clearly confirm that it is extremely difficult to verify realized savings of past multiyear programs or to accurately predict savings on future MYP contracts. Yet, in spite of these challenges, our survey of the literature and analysis of historical MYP contracts demonstrate the following:

- There are significant differences among the key elements characterizing historical multiyear aircraft programs. Each program must be evaluated in depth and on its own terms with respect to its unique characteristics. Notwithstanding this great diversity, there is a consensus among most program officials and subject-matter experts that EOQ and CRIs are important drivers of savings on most MYPs.

- Comprehensive and in-depth analysis of contractor cost structure beyond what has typically been done in the past, particularly on the lower tiers, can substantially increase estimated program savings, particularly during the contract negotiation phase.

- Development of an appropriate baseline estimate of the likely cost of comparable single-year contracts is difficult but crucial as the baseline against which government-estimated MYP savings will ultimately be determined.

The remaining sections in this appendix provide considerably greater detail on the more recent case studies we examined. These provide greater insight and context about the MYP and BB contracts that we evaluated and further illustrate the broad diversity across the historical BB and MYP contracts. Each case also notes general similarities and differences between it and the proposed F-35 BB program.

---

[22] The more detailed discussion of our MYP case studies include very high-level quantitative analyses based on SAR data that provide insights into whether an MYP achieves the cost targets established by the contract, but still does not indicate how much savings are achieved by attaining the target price that was negotiated compared with comparable single-year contracts.

[23] Many academic and government studies over the past three decades have recognized these problems. See, for example, V. Sagar Bakhshi and Arthur J. Mandler, *Multiyear Cost Modeling*, Fort Lee, Va.: Army Procurement Research Office, Office of the Deputy Chief of Staff for Logistics, APRO 84-03, February 1985; GAO, *DoD's Practices and Processes for Multiyear Procurement Should Be Improved*, Washington, D.C., GAO-08-298, February 2008; and O'Rourke and Schwartz, 2015.

C-17A

The C-17A, or Globemaster III, is a strategic airlift aircraft operated by the U.S. Air Force and eight other international customers.[24] Operational since 1995, the C-17A has a maximum cargo capacity of 585,000 pounds (or 101 passengers, or 18 pallets) and is capable of taking off from—and making steeper, slower landings on—shorter, austere runways. In addition to air-landing cargo, the platform is capable of air-dropping a wide variety of equipment as well as personnel. Boeing's Long Beach facility in California represented the main hub of the C-17A's development as well as production. Key subcontractors included Pratt & Whitney (P&W), which produced the engines, and L-3 Communications, which manufactured training systems.

Although the original procurement plan envisioned 120 aircraft, the Air Force eventually purchased a total of 224 C-17As.[25] Industrial base concerns, shifting strategic priorities, and international sales helped sustain the C-17A production line in its latter years. Although Boeing delivered the last Air Force C-17A in September 2013, it continued to produce C-17As for foreign customers. Out of the 267 delivered worldwide, 44 of those aircraft were for its eight international customers through Foreign Military Sales (FMS) and direct commercial sales. International customers include Canada (5), Kuwait (2), India (10), Qatar (4), Australia (6), United Arab Emirates (6), United Kingdom (UK) (8), and North Atlantic Treaty Organization (3).

The C-17 program used two MYP contracts during its procurement; the first one was the largest and longest MYP contract signed at the time. Combined, the MYP contracts projected to save $2.4 billion over the use of single-year contracts.

Characteristics Prior to MYP

Before the award of the C-17's first MYP contract, the program was just emerging from low-rate production. Prior production contracts initially consisted of FPIF contracts, followed by FFP contracts for the final two annual contracts before the MYP contract's award. All eight lots, however, had fully certified cost or pricing data available to the program office to assist with contract negotiations.[26] Altogether, the C-17 program had produced 32 aircraft, with an annual production rate only reaching up to six aircraft a year (see Table B.4).

Early in 1996, the Air Force requested approval for a seven-year MYP contract for the C-17 program, estimating a savings of about $896 million, or approximately 5 percent. In addition to

---

[24] Elements of this section were derived from research by RAND colleagues William Shelton, Stephen Joplin, Cynthia Cook, Abby Doll, James Dryden, Bernard Fox, Mark Lorell, Karishma Mehta, Leslie Payne, Katherine Pfrommer, Soumen Saha, and Cole Sutera.

[25] As of May 2013, one delivered Air Force C-17 had been retired and one had been destroyed in a Class A mishap. Thus, at that time, the Air Force owned 223 C-17s, but operated 222.

[26] Mark A. Lorell, John C. Graser, and Cynthia Cook, *Price-Based Acquisition: Issues and Challenges for Defense Department Procurement of Weapon Systems*, Santa Monica, Calif.: RAND Corporation, MG-337-AF, 2005.

breaking the norm of MYP contracts lasting only five years, the MYP contracts eventually signed with McDonnell Douglas and P&W for the engine also became the largest MYP effort at the time, totaling $16.2 billion. Despite the projected savings, GAO expressed several doubts about whether the C-17 met all of the statutory requirements for an MYP contract. Raising design stability concerns, the GAO report noted that 95 items remained open at the time on the Initial Operational Test and Evaluation report that could lead to changes in the aircraft's design. In addition, test and evaluation of all required performance specifications was not expected for another 18 months. Also, the Air Force projected a future need for an additional $1 billion in research and development funds for future enhancements, $275 million for engineering change orders, $308 million for product improvements, and $1.2 billion for modifications. The GAO report also highlighted unresolved differences regarding the number of C-17s required for the overall airlift mission, potentially threatening future quantity stability. Finally, because of ongoing affordability initiatives and the contractor's characterization of a large portion of MYP savings as "management challenges," the GAO report questioned why savings projected under the MYP contract could not be accomplished under single-year procurements.[27] Congress ultimately supported the MYP buy under the conditions that the contract require additional savings beyond 5 percent and allow for a return to single-year procurement without penalty.[28]

Table B.4. Characteristics of C-17 Prior to MYP Award

Characteristic	Contract (Award Year)	Type	Quantity	Cost and Pricing Data Provided
Production contracts	Lot 1 (1988)	FPIF	2	Y
	Lot 2 (1989)	FPIF	4	Y
	Lot 3 (1990)	FPIF	4	Y
	Lot 4 (1991)	FPIF	4	Y
	Lot 5 (1993)	FPIF	6	Y
	Lot 6 (1994)	FPIF	6	Y
	Lot 7 (1994)	FFP	6	Y
	Lot 8 (1996)	FFP	8	Y
GAO concerns about MYP eligibility	• Severe cost overruns in recent years almost leading to cancellation • Design deficiencies • Potential quantity instability due to unstable airlift requirements • Ongoing cost reduction efforts already producing savings under single-year contracts			

---

[27] GAO, 1996.

[28] Sheila Foote, "Appropriators Back C-17 Multiyear Buy, with Conditions," *Defense Daily*, April 1, 1996a; Sheila Foote, "Senate Passes Bill Authorizing C-17 Multiyear Buy," *Defense Daily*, May 24, 1996b.

Award of First MYP Contract

In 1996, DoD awarded McDonnell Douglas an FFP MYP contract for 80 aircraft. The program office estimated a savings of around $1.2 billion, or 5.5 percent, over the seven-year period (see Table B.5).[29] As noted above, McDonnell Douglas originally offered savings of 5 percent, but pressure from Congress led the contractor to add 0.5 percent as a "management challenge." The MYP effort included $100 million in CRI funding, as well as $300 million in EOQ funding expected to return $900 million in savings.[30] Before the contract's final negotiation, Air Force officials estimated that $100 million of the EOQ funding would be applied to affordability projects, although it is unclear whether this was executed as stated.[31] An interesting aspect of the contract was a clause permitting return to a contract with single-year options without paying cancellation costs if a production lot under the MYP contract was not fully funded.

CRI funding was considered crucial in producing savings, although a large amount of CRI funding occurred before the award of the first MYP and overlapped across contract awards. Before the award of the first MYP, according to budget documents, EOQ funding was also available for use in CRI efforts. Altogether, it is estimated that approximately $560 million in CRI investments either occurred during or overlapped with the first MYP contracts, not including as much as $100 million in EOQ funding that may have been diverted to CRIs.[32]

Award of Second MYP Contract

The second $9.8 billion MYP contract, also FFP, was awarded in 2002 for 60 aircraft and for an estimated savings of $1.3 billion, or 10 percent.[33] The MYP contract was derived from an unsolicited, nine-page proposal from Boeing that did not include TINA-certified cost and pricing data and instead used a price-based acquisition approach.[34] The government did not provide CRI funding, although EOQ made up 7 percent of the total contract value, allocated mostly during the first years of the contract. As an FFP contract, it did not require Earned Value Management (EVM) data reporting. Other interesting aspects of this contract are that:

---

[29] It is unclear whether the 5.5 percent arose from only the air vehicle production contract or if the estimate also includes the engine program, which also gained an MYP contract for $1.6 billion at the same time. This discrepancy is described in further detail in Younossi et al., 2007. See also DoD, *C-17A Globemaster III*, Selected Acquisitions Reports, executive summaries, assorted dates.

[30] Bruce R. Harmon, Scot A. Arnold, James A. Myers, J. Richard Nelson, John R. Hiller, M. Michael Metcalf, Harold S. Balban, and Harley A. Cloud, "F-22A Multiyear Procurement Business Case Analysis," Institute for Defense Analyses, P-4116, undated (circa 2006).

[31] GAO, 1996.

[32] Younossi et al., 2007.

[33] DoD, *Selected Acquisition Report: CH-47F Improved Cargo Helicopter*, 2003.

[34] Price-based acquisition typically employed an FFP contract with the price established without the use of detailed certified TINA–supplier cost data. Office of the Under Secretary of Defense for Acquisition, Technology, and Logistics, *Report of the Price-Based Acquisition Study Group*, November 15, 1999.

- It was originally intended to be a commercial-item, FAR Part 12 contract but did not receive authorization. However, the contract did include TINA waivers and used a price-based acquisition approach.[35]
- Additional savings were sought through the use of incremental funding and a larger cancellation liability, which eventually reached an estimated $1.5 billion.[36] Congress became concerned that this practice violated DoD's full-funding policy and the Antideficiency Act and subsequently increased procurement funding in fiscal years (FYs) 2003 and 2005 and prohibited incremental funding of MYP contracts.[37]

As with the first MYP contract, increased indexes for EPA led to estimated funding shortfalls of $185 million in FY 2006 and $158 million in FY 2007. Negotiations with the contractor, however, led to an offer executable within the existing budget.[38] With no cost reporting, it was difficult for the government to track the exact use of EOQ funding and the sources of cost savings. Boeing asserted that it was able to negotiate savings of 10–20 percent through vendor discounts, although it is unclear whether it was achieved through EOQ funding.[39]

Table B.5. Comparison of C-17 MYP I and MYP II

Characteristic	MYP I	MYP II
Period of performance	FYs 1997–2003 (lots 9–15)	FYs 2003–2007 (lots 15–19)[a]
Contract amount at award (TY $)	$14.2 billion (1996)	$9.8 billion (2002)
Last reported contract amount (TY $)	$16.6 billion (2003)	$10.0 billion (2007)
Quantities	80	60
Build rate	8, 9, 13, 15, 12, 15, 8	7, 11, 15, 15, 12
Contract type	FFP	FFP
CRI funding	$0[b]	$0[b]
EOQ funding (TY $)	$300 million (began with lot 8)	$645 million
Estimated savings (TY $)	$1.2 billion (5.5%)	$1.3 billion (10.8%)
Cost and pricing data regularly reported	Until lot 12	No

[a] Lot 15 was split between MYP I and MYP II (lots 15 and 15B). Lot 15 ended and lot 15B began in the same fiscal year.
[b] CRIs were not associated with the particular MYP contract; rather, they were incorporated through other, single-year contracts.

---

[35] Younossi et al., 2007.

[36] Due to lack of clarity in the FARs, the program office sought to apply advanced procurement (AP) funds (EOQ funding) for future aircraft that were not yet fully authorized in the budget, but were expected to be later. This was an attempt to achieve EOQ savings before the anticipated future aircraft were fully authorized by Congress.

[37] GAO, 2008.

[38] Defense Acquisition Executive Summary Report, *Assessments for the February 2006 Review*, summary for C-17A, PNO 200, February 2006a; Defense Acquisition Executive Summary Report, *Assessments for the May 2006 Review*, summary for C-17A, PNO 200, May 2006b.

[39] Younossi et al., 2007.

In the C-17 MYPs, internally funded, contractor-reported savings efforts, including aggressive lean efforts and management challenges, appear in contractor-provided data to have achieved touch labor savings.[40] Both the contractor and the program office described the C-17 program as a "survival story," where the threat of cancellation or decreased buys helped to incentivize savings. Discussions with Boeing also suggested that having a single program site contributed to this urgency. Although Boeing was making foreign sales, it faced competition from Airbus for overseas contracts. Boeing also had made investments toward the award of the Joint Strike Fighter contract, and when this did not materialize, Boeing had to reevaluate its corporate strategy.

The F-35 program, as it stands, does not face high risk of total program cancellation. Therefore, the "survival" forces attributed to driving down C-17 costs are not necessarily applicable. The first MYP for the C-17 is also unique among all the programs explored in this appendix by virtue of its length (seven years instead of the typical five) and the option to return to single-year procurement without penalty. Despite these differences, the C-17 case does potentially show the value of CRI funding outside of an MYP contract, particularly early in the program.

Another interesting aspect of this program is that it illustrates the lack of rigor often evident in past MYP contracts regarding the congressional application and assessment of the statutory requirements for MYP approval. GAO argued that for C-17 MYP I, many of the mandatory requirements (particularly regarding a stable and mature design) did not seem to be rigorously applied. Political and economic factors obviously affected Congress's decision. This may partially explain why BBs have been used so rarely in the past.

F/A-18E/F

Designed to replace the F/A-18C/D, A-6, and F-14, the F/A-18E/F Super Hornet is a twin engine, midwing, multimission tactical aircraft that performs both air-to-air and air-to-ground missions, including fighter escort, interdiction, air defense, close air support, and aerial tanking.[41] In addition to the one- and two-seat variants, the program includes the EA-18G, introduced in FY 2006, which is an electronic warfare variant known as the Growler. Although the Super Hornet looks similar to its predecessor, the Hornet, it has a longer fuselage and a 25-percent larger wing. As the prime contractor, Boeing performs the final assembly, as well as the manufacture of the forward fuselage in St. Louis, Missouri. Key subcontractors include

---

[40] Cost reduction projects by Boeing included process changes, component/system redesign, work transfer/second sourcing, material changes, tooling changing, quality improvements, and facilities utilization.

[41] Elements of this section were taken from research by RAND colleagues William Shelton, Stephen Joplin, Cynthia Cook, Abby Doll, James Dryden, Bernard Fox, Mark Lorell, Karishma Mehta, Leslie Payne, Katherine Pfrommer, Soumen Saha, and Cole Sutera.

Northrop Grumman, which produces the aft and center fuselage sections and vertical tails (and integrates all associated subsystems) in El Segundo, California. General Electric produces the engines, and flight testing is conducted by the Naval Air Warfare Center in Patuxent River. Although the F/A-18E/F was approved for FMS in 2001, only the Australian Air Force has procured it, with a total of 24 aircraft in its inventory.[42] Overall, the F/A-18E/F program has involved three MYP contracts that have spanned the majority of its U.S. procurement.

Characteristics Prior to MYP

Much of the F/A-18E/F's acquisition approach derived from lessons learned by the cancellation of the A-12, which was developed by McDonnell Douglas/General Dynamics and ultimately canceled in 1991 because of cost and schedule overruns (see Table B.6).[43] Informed by the "Hornet 2000" studies of the 1980s, the Navy chose to pursue a low-risk option to upgrade the F/A-18C/D capability set. On May 12, 1992, the Undersecretary of Defense approved the Navy's request to designate the F/A-18E/F's development as a major modification to the F/A-18C/D instead of a new build, primarily because the program was considered to be low risk. By maintaining a similar contractor team structure as with the F/A-18C/D, McDonnell Douglas and Northrop were able to draw from existing subcontractors and design teams. The F/A-18E/F was also chosen as the "proof of concept" for the emerging integrated product teams concept, which also contributed to meeting schedule and cost goals overall.[44] In 1997, McDonnell Douglas merged with Boeing, which inherited the F/A-18E/F program. Despite efforts to reduce risk, the GAO expressed concerns about design stability, because 71 deficiencies from the development phase were still not corrected as of the report's release in June 1999.[45] Before the award of the first MYP, the F/A-18E/F program had two production contracts for LRIP, as shown in Table B.6. The first was a cost plus incentive fee (CPIF) contract for 12 aircraft, and the second was an FPIF contract for 50 aircraft.

Table B.6. Characteristics of F/A-18E/F Prior to MYP Award

Characteristic	Contract (Award Year)	Type	Quantity	Cost and Pricing Data Provided
Production contracts	LRIP I (1996)	CPIF	12	Y (assumed)
	LRIP II/III (1997)	FPIF	50	Y (assumed)
Concerns about MYP eligibility	• Design deficiencies			

---

[42] DoD, *Selected Acquisition Report: F/A-18E/F Super Hornet Aircraft (F/A-18E/F)*, December 2001a.

[43] Obaid Younossi, David E. Stem, Mark A. Lorell, and Frances M. Lussier, *Lessons Learned from the F-22 and F/A-18E/F Development Programs*, Santa Monica, Calif.: RAND Corporation, MG-276-AF, 2005.

[44] Younossi et al., 2005.

[45] GAO, *Defense Acquisitions: Progress of the F/A-18E/F Engineering and Manufacturing Development Program*, Washington, D.C., GAO/NSIAD-99-127, June 1999.

Award of First MYP Contract

By June 2000, the F/A-18E/F program entered FRP with the signing of a five-year MYP FPIF contract for 222 aircraft (later decreased to 210). An FPIF contract, with a 70:30 profit share–line split, was chosen specifically because the aircraft was just entering FRP and design deficiencies remained unaddressed.[46] A total of $200 million was provided for EOQ, although the F/A-18E/F cost reduction team received permission from the Navy to use this funding as "nonrecurring producibility improvement funds," as long as they were used for supplier initiatives. Altogether, the MYP justification package estimated that the MYP contract would achieve savings of $706 million (7.4 percent) over single-year procurement. Vendor procurement totaled 68 percent of the projected savings. CRIs and quantity discounts acquired because of the large procurement numbers, however, were later asserted to be the major source of savings with suppliers rather than EOQ funding.[47] Naval Air Systems Command, as well as Boeing officials, asserted that the majority of realized savings came primarily from CRI funding, which also flowed down to vendors. Other interesting aspects of the contract and its execution include the following:

- A variation in quantity clause allowed for an increase or decrease of six aircraft per year.
- In May 2002, the program reported that EPA increases were unfunded, which were attributed to new union contracts and rate adjustments as a result of Boeing losing the F-35 bid. Altogether for the first MYP contract, the government paid $378 million in unexpected price adjustments, and the GAO found a 10-percent increase in average unit costs compared with the original budget estimates.

Award of Second MYP Contract

In December 2003, the Navy awarded Boeing a second MYP contract for 222 aircraft (later reduced to 179), which, in contrast to the first MYP, used an FFP contract and provided $100 million for CRIs but no funding for EOQ. Using price-based acquisition, the contract provided TINA waivers and reduced reporting and oversight requirements, in an attempt to emulate commercial-style practices.[48] The second MYP contract also saw the introduction of the EA-18G, which constituted 56 of the total aircraft buy. Program officials assert that their analysis of the first MYP from the previous contract revealed that CRIs yielded significantly more savings than EOQ; after polling its subcontractors, Boeing agreed with this assessment.[49] Thus,

---

[46] GAO, 2008. Thus, if the price rose above the target price toward the ceiling price, the government would take responsibility for 70 percent of the additional cost, while the contractor only had to incur 30 percent of the additional cost. This was relatively generous for the contractor because of existing design uncertainties.

[47] Younossi et al., 2007.

[48] Younossi et al., 2007.

[49] Discussions with program officials, December 2015.

no EOQ was offered in either MYP II (or III). Other interesting aspects of the contract include the following:

- The buy could be increased by up to six aircraft; however, the original negotiated quantities could not be decreased. The program manager noted, "given historical funding shortfalls, this is a high-risk strategy."[50]
- Officials worried that potential unbudgeted future cost increases from the EPA might require offsets in other programs' funding in later years.[51] This, in addition to Boeing's unexpected requirement for a large pension fund contribution, led to an obligation by the Navy to pay more than $1 billion; negotiations, however, brought the price adjustment down to $152 million.[52] Increased indexes in aluminum and titanium contributed greatly to the price increase. Program officials speculated that Boeing could deliberately affect the index because of its sheer presence in the market.[53]

Award of Third MYP Contract

The third MYP contract, awarded in December 2008, was initially intended to procure 66 F/A-18E/Fs (later increased to 103) and 58 EA-18Gs, expected to produce $820 million in savings over single-year contracts, and was FPIF.[54] An FPIF contract was specifically chosen because of what was perceived to be excessive profit for Boeing and Northrop Grumman during MYP II.[55] Other interesting aspects of the contract include the following:

- Negotiations originally resulted in pricing based on 151 aircraft, although only 124 were initially funded. It was assumed that additional aircraft would be added later.[56]
- In 2010, a Defense Federal Acquisition Regulation Supplement, Subpart 234.2 Individual Deviation, request was approved to omit EVM requirements from the contract.[57]

---

[50] Defense Acquisition Executive Summary Report, *Assessments for the January 2004 Review*, summary for F/A-18E/F, PNO 549, January 2004a.

[51] Defense Acquisition Executive Summary Report, *Assessments for the January 2005 Review*, summary for F/A-18E/F, PNO 549, January 2005a.

[52] GAO, 2008.

[53] Discussions with program office officials, December 2015.

[54] Dan Taylor, "Navy Awards $5.3 Billion Contract to Boeing for F/A-18, EA-18G Multiyear," *Inside the Navy*, October 4, 2010.

[55] Program officials estimate that Boeing's profit percentage points were in the high 20s and lower 30s, while Northrop Grumman reached the high 20s in its profit percentage. Discussions with program officials, December 2015.

[56] Discussions with the program office, December 2015.

[57] Defense Acquisition Executive Summary Report, *Assessments for the April 2013 Review*, summary for F/A-18E/F, PNO 549, April 2013b.

The three F/A-18E/F MYPs are summarized in Table B.7. In the F/A-18E/F program, program officials applied lessons learned from each previous MYP contract while negotiating the next. The second MYP entailed a "hands-off," more commercial-like approach, with little in-depth tracking of how Boeing spent CRI funding; the third MYP, however, saw more-active program office participation in the CRI vetting and approval process.[58] As with the C-17 program, the contractor reported significant savings as a result of its CRI efforts, which included aggressive lean initiatives and supplier engagement and challenges.[59] Although its design was more stable and less complex, the F/A-18E/F program is one of the closer analogs to the F-35 program. In particular, an important lesson is the value of intensive analysis of outcomes and cost data from the previous contracts to achieve greater leverage, and thus greater savings, for subsequent contracts.

Table B.7. Comparison of F/A-18E/F MYP I–MYP III

Characteristic	MYP I	MYP II	MYP III
Period of performance	FYs 2000–2004	FYs 2005–2009	FYs 2009–2013
Contract amount at award (TY $)	$9.0 billion (1999)	$8.7 billion (2004) (*$6.4 billion (F/A-18E/F)* + *$2.2 billion (EA-18G)*)	$5.3 billion (*$2.8 billion (F/A-18E/F)* + *$2.5 billion (EA-18G)*)
Last reported contract amount (TY $)	$9.2 billion (2006)	$10.3 billion (2011) (*$7.7 billion (F/A-18E/F)* + *$2.5 billion (EA-18G)*)	$7.2 billion (2012 F/A-18E/F, 2015 EA-18G) (*$4.5 billion (F/A-18E/F)* + *$2.6 billion (EA-18G)*)
Quantities	222 F/A-18E/F (later reduced to 210)	154 F/A-18E/F, 56 EA-18G (later increased to 179 F/A-18E/F)	66 F/A-18E/F, 58 EA-18G (later increased to 103 F/A-18E/F)
Contract type	FPIF (70:30)	FPEPA (FFP)	FPIF
CRI funding	$115 million	$100 million	$100 million
EOQ funding	$85 million	$0	$0
Estimated savings	$706 million (7.4%)	$1.052 billion (10.95%)	$818.8 million (10%)
Cost and pricing data regularly reported	Yes (assumed)	Yes (assumed)	Yes (assumed)

NOTE: FPEPA = fixed price with economic price adjustment.

Unlike nearly all recent MYPs, the F/A-18E/F program focused much more heavily on CRIs compared with EOQ as the most significant source of savings. Indeed, the second and third MYPs included no EOQ funding. This runs contrary to the views of many officials regarding the importance of EOQ funding for savings, although this view is shared by other programs, such as

---

[58] Discussions with program officials, December 2015.

[59] Discussions with Boeing, 2014.

the CH-47F. It seems intuitively implausible that EOQ funding would have no effect on MYP savings, but we could find no statistically significant quantitative evidence proving it. While most officials believe EOQ funding is very important for savings, this was not the case on the second two F/A-18E/F programs or on the CH-47F. Unfortunately, the unique characteristics of each MYP, the many changes to the original plan that take place during MYPs, and the lack of relevant metrics and data collection, make resolution of this issue extremely difficult if not impossible. But it is important to note that programs that included no EOQ funding (and in some cases no CRI funding) were nonetheless estimated to achieve significant savings. A good example of this is the CH-47F MYP II, which included neither EOQ funding nor CRI funding yet achieved an estimated savings of nearly 20 percent once the prime contract was finalized, following the implementation of a careful negotiation strategy.

C-130J

The C-130J has many military variants that range across mission areas, as well as a commercial variant, the LM-100J.[60] The C-130J is designed primarily for transport of cargo and personnel within a theater of operations, although 11 variants of the C-130J are capable of performing other types of missions, including rescue and recovery, air refueling, special operations, firefighting, and weather reconnaissance. Aside from the Air Force, the Marine Corps operates the KC-130J variant and the Coast Guard operates the HC-130J variant. Lockheed Martin Aeronautics (LMA), as the prime contractor, performed the design and final assembly at its facility in Marietta, Georgia. Key subcontractors include Rolls Royce (RR), which produces the engines in Indianapolis, Indiana; Dowty, which produces the propellers in the UK; and Lockheed Martin Integrated Systems and CAE,[61] which produces the training systems in Orlando and Tampa, Florida. Sustainment is performed by Warner Robins Logistics Center at Robins Air Force base in Georgia, and flight-testing is conducted by the 412th Test Wing at Edwards Air Force Base in California.[62]

Characteristics Prior to MYP

After initial market analysis and discussions with the Air Force in 1989, the C-130J began as a privately funded development and test program by LMA in 1991. LMA sought an initial operating capability by September 1994 to meet a military airlift command requirement for enhanced theater

---

[60] Elements of this section were derived from derived from research by RAND colleagues William Shelton, Stephen Joplin, Cynthia Cook, Abby Doll, James Dryden, Bernard Fox, Mark Lorell, Karishma Mehta, Leslie Payne, Katherine Pfrommer, Soumen Saha, and Cole Sutera.

[61] This company was formerly Canadian Aviation Electronics.

[62] Timrek Heisler, *C-130 Hercules: Background, Sustainment, Modernization, Issues for Congress*, Washington, D.C.: Congressional Research Service, R43618, June 24, 2014.

airlift, and it was reported that investments needed to start in 1989 to meet that goal.[63] By 1993, the internally funded development program had progressed to the production design phase, with a first flight projected to occur in September 1995. Nonetheless, the Air Force did not include additional C-130s in its FY 1994 budget request.[64] Because the program was contractor funded and Air Force did not provide funds for nonrecurring costs associated with development, the program never had a specific Operational Requirements Document or an LRIP phase.[65] Because of other budget priorities, the Air Force continued to delay requesting procurement funding for C-130Js until the early 2000s. Instead, for a variety of political and economic reasons, Congress authorized funding for around 20 of the aircraft between 1994 and 1998.[66] The Air Force even declined a $20 million savings offer from LMA to transfer ownership of completed C-130Js early in 1998.[67] The Air Force began requesting funding for C-130J procurement in earnest beginning in 2002, when the first discussions for an MYP contract began.

Award of First MYP Contract

When negotiations for an MYP contract began in 2002, Congress expressed concern about potentially unfunded activities in the MYP, peaking in FY 2004 at $300 million. LMA would have to borrow money that DoD would subsequently have to reimburse, in addition to any interest accrued. The primary concern was that of unfunded termination liability cost should the program be discontinued before the completion of the MYP contract.[68] See Table B.8 for characteristics of the program prior to the first MYP.

The restructured MYP, as published in the PB, sought to ease these concerns through "slowed production and [reduction in] the unfunded cancellation ceiling by increasing funding for Economic Order Quantity and adding funding for advance procurement." The PB states that, "even with the restructure, an unfunded cancellation liability of $40.5 million still exists in FY 2004, and a total of $120.8 million through FY 2007 . . . although the unfunded amount is now below the threshold of 20 percent established in the Department's Financial Management

---

[63] Edward H. Kolcum, "Lockheed Weighs Investment Risks of Developing New C-130 Version," *Aviation Week & Space Technology*, Vol. 131, No. 22, November 27, 1989.

[64] "C-130J Upgrade Targets Aging Transport Fleets," *Aviation Week & Space Technology*, Vol. 139, No. 2, July 12, 1993.

[65] As of June 1999, no Air Force funds had been used for nonrecurring costs associated with development. "Lockheed Officials Say C-130J Upgrades Are Effective and on Schedule," *Inside the Air Force*, June 11, 1999.

[66] Brendan Sobie, "Air Force to Postpone High-Rate C-130J Acquisition Until at Least FY-03," *Inside the Air Force*, April 10, 1998.

[67] This offer was made in an effort to move the aircraft off the "liability ledger" and show better profits for 1998. Adam J. Hebert, "Air Force Declines Lockheed Offer for Conditional Acceptance of C-130J," *Inside the Air Force*, January 8, 1999.

[68] "Amid Criticism, Air Force Says C-130J Multiyear Funding Profile Is Sound," *Inside the Air Force*, June 21, 2002.

Regulation."[69] Also, the MYP assumptions on cost reduction depended on the Coast Guard or international customers to procure aircraft to maintain an efficient production rate of 16 aircraft per year.[70]

Table B.8. Characteristics of C-130J Program Prior to MYP Award

Characteristic	Contract (Award Year)	Type	Quantity	Cost and Pricing Data Provided
Production contracts before MYP I	Production (1996)	FFP	19	No
	Production (2000)	FFP	19	No
	Five-Year Option Contract III	FFP	42	No
Production contracts before MYP II	Five-Year Option Contract IV	FFP	2	No
Concerns about MYP eligibility	• Limited visibility into technological, design, or production maturity due to commercial nature of program • Limited visibility into cost and pricing data from prior single-year contracts • Shifting requirements for strategic lift capabilities			

The FFP MYP contract was awarded on March 14, 2003. Further criticism arose in a 2004 DoD Inspector General report that asserted the C-130Js procured to date did not meet operational and contractual requirements, yet the Air Force paid 99 percent of the contract price to LMA, "leaving the contractor little financial incentive to correct deficiencies."[71] In addition to not meeting operational requirements, C-130J defects were found to increase operations and maintenance costs. The report also stated that the commercial item acquisition strategy was unwarranted because it did not meet federal acquisition regulations that require the product to be intended customarily for commercial use, need only minor modifications to meet government requirements, and be available to the commercial market for sale. The H-model was only sold to governments, and no true commercial variant existed at the time of the contract. The use of a commercial contract vehicle also prevented access by DoD to contractor cost, pricing, and profit data.[72]

The FY 2006 PB projected cancellation of the Air Force program that fiscal year, followed by the Marine Corps program the following year. However, program officials noted, "There are numerous inconsistencies and apparent conflicts in the contract terms and conditions applicable to termination for convenience and/or multiyear contract cancellation that will require extensive

---

[69] "Lockheed Martin Snags $4 Billion Contract for C-130J Multiyear Deal," *Inside the Air Force*, March 21, 2003.

[70] "Lockheed Martin Snags $4 Billion Contract for C-130J Multiyear Deal," 2003.

[71] DoD Office of the Inspector General, *Acquisition: Contracting for and Performance of the C-130J Aircraft*, Washington, D.C., D-2004-102, July 23, 2004, p.3.

[72] DoD Office of the Inspector General, 2004.

legal review and may significantly impact negotiation of a termination settlement."[73] Estimates of the termination cost rose from \$439.7 million to \$2 billion.[74] In May 2005, DoD decided against cancelling the C-130J and moved forward with the contract conversion from FAR Part 12 to FAR Part 15, or from a commercial-item procurement to a traditional military procurement. The MYP was converted into a traditional FAR Part 15 defense contract on October 18, 2006. On December 15, 2006, LMA was awarded a contract for three Air Force and one Marine Corps aircraft funded in the FY 2006 Global War on Terror Supplemental on the Five Year Option Contract III because all ordering slots on the restructured MYP were filled.[75] In November 2005, there was "agreement in principle" between the Air Force and LMA to apply TINA compliance with future pricing actions after the contract's conversion to a FAR Part 15 contract.[76] Yet despite the conversion to a traditional noncommercial contract, program officials asserted that they still did not have visibility into production metrics and instead relied on DCMA representatives at the production facility to ensure standards and proper documentation.[77]

Other interesting aspects of the contract include the following:

- The contractor initially agreed to maintain a workforce to produce 16 aircraft per year, although DoD was funded to purchase fewer aircraft during several years. If the balance was not made up by foreign military or direct commercial sales, the contract included a provision to adjust prices retroactively.[78] On September 28, 2005, however, the price re-opening terms were eliminated in the contract.

- Supplier reporting initially was not to be required after the conversion. However, with input from Congress, the flow of FAR Part 15 provisions to subcontractors was eventually included.[79]

- To help with developing projected MYP savings, A. T. Kearney reportedly spent four months observing the assembly and flight lines to accumulate observed actuals. It also evaluated several potential cost-reduction projects and showed estimated hours-per-unit

---

[73] Defense Acquisition Executive Summary Report, *Assessments for the May 2005 Review*, summary for C-130J, PNO 220, May 2005b.

[74] "Pentagon Plans to Terminate C-130J Procurement in FY-06 Budget," *Inside the Air Force*, January 7, 2005; "Senior USAF Officer: C-130J Termination Costs Could Exceed \$2 Billion," *Inside the Air Force*, March 4, 2005.

[75] DoD, *Selected Acquisition Report: C-130J Hercules Transport Aircraft (C-130J)*, December 2006a.

[76] Defense Acquisition Executive Summary Report, *Assessments for the November 2005 Review*, summary for C-130J, PNO 220, November 2005d.

[77] GAO, *Defense Acquisitions: Assessments of Selected Weapon Programs*, Washington, D.C., GAO-07-406SP, March 2007.

[78] Defense Acquisition Executive Summary Report, *Assessments for the May 2003 Review*, summary for C-130J, PNO 220, May 2003.

[79] Defense Acquisition Executive Summary Report, *Assessments for the August 2006 Review*, summary for C-130J, PNO 220, August 2006c.

savings. To the program office's knowledge, however, none of the projects were instituted.[80]

The Air Force claimed that the conversion produced an additional net savings of $167.7 million as the result of the use of TINA-compliant cost and pricing data used to renegotiate the remaining aircrafts' prices.[81] Acquisition officials claim that LMA successfully negotiated supplier prices down about 13 percent on average for a FY 2003–2008 buy.[82]

Award of Second MYP Contract

The award of the second MYP in 2016 was significantly delayed because of inadequate proposals and slow response times from LMA during negotiations, described by the program manager as "a longstanding, systematic challenge for the program."[83] An Undefinitized Contract Action (UCA) provided MYP II EOQ funding, in addition to a second UCA for an FY 2013 congressional add-on (of ten aircraft) in December 2014.[84] Budget documents submitted in 2014 outlined a $6 billion MYP buy of 83 C-130Js for the Marine Corps, Coast Guard, Air Force, and Special Operations Command, a buy estimated to save 9.5 percent over annual contracting. The final $5.3 billion contract for 78 aircraft was announced in December 2015, with the Office of the Secretary of Defense (OSD) Cost Assessment and Program Evaluation (CAPE) savings projection of 11.5 percent.[85] The source of the projected savings was not clear at the time this report was written.

Summary and Lessons for F-35 BB

Table B.9 compares the two C-17 MYPs. Because of the late transition from a commercial to a military procurement, program officials remained uncertain of the true magnitude and sources of MYP savings.[86] A clear picture of what exact efforts were implemented to achieve savings also was not available at the time this report was written. Because of the unique trajectory the program took, it is difficult to draw direct applicable lessons for the F-35 program. The transition from a FAR Part 12 to a FAR Part 15 procurement, however, does reveal the benefits of

---

[80] Discussions with C-130J program office, January 2015.

[81] Department of Defense Office of the Inspector General, *Review of Defense Contract Management Agency Support of the C-130J Aircraft Program*, Report No. D-2009-074, June 12, 2009.

[82] Younossi et al., 2007.

[83] Defense Acquisition Executive Summary Report, *Assessments for the February 2013 Review*, summary for C-130J, PNO 220, February 2013a.

[84] Defense Acquisition Executive Summary Report, *Assessments for the February 2015 Review*, summary for C-130J, PNO 220, February 2015.

[85] DoD, 2015. See also Lockheed Martin Aeronautics Company, "U.S. Government, Lockheed Martin Announce C-130J Super Hercules Multiyear II Contract," *PR Newswire*, December 31, 2015.

[86] Discussions with C-130J program office, December 2015.

increased visibility into historical costing data in negotiation leverage. Although it is unclear whether any of the suggestions were implemented, A. T. Kearney's role as a "surrogate" analytic arm to help determine further savings could also add to this leverage. This supports the arguments made earlier regarding the importance of obtaining detailed cost knowledge, especially regarding lower-tier contractors, prior to final contract negotiations.

Table B.9. Comparison of C-130J MYP I and MYP II

Characteristic	MYP I	MYP II
Period of performance	FYs 2003–2008	FYs 2017–2021
Contract amount	$4.05 billion	$5.3 billion reported in MYP justification, $1.8 billion at contract signing
Quantities	60	78 reported in MYP justification, 29 at contract signing
Contract type	FFP	FPIF
CRI funding	None identified (EOQ may have been used)	None identified
EOQ Funding	$140 million	$92.0 million[a]
Estimated savings	$324 million (10.9%) for Air Force and $177 million (9.8%) for Marine Corps	11.5%[b]
Cost and pricing data regularly reported	No, but limited after 2006.	Yes (assumed)

[a] From original 2014 budget documentation.
[b] From 2015 C-130J SAR.

F-22A

Designated the "Raptor," the F-22A is a single-seat, fifth-generation air platform incorporating low observable technology, supersonic cruise without afterburner, and internal weapons and fuel storage. The F-22A emerged from requirements for the Advanced Tactical Fighter to replace the F-15 in the air superiority mission. In 1991, the Air Force awarded development contracts to the LMA-Boeing team for the airframe and to P&W for the F119 engine. The addition of the air-to-ground role to the platform's requirements led to its designation as the F/A-22 and finally in 2005, the F-22A. Like the proposed F-35 BB, the F-22A program had a shorter MYP contract of three years. Unlike the F-35 BB, however, the three-year MYP took place at the end of F-22 production.

Characteristics Prior to MYP

Before the award of its MYP contract, the F-22A program experienced cost and schedule growth, as well as development challenges that led to increased scrutiny by DoD and Congress.

Design issues included loading on the tail boom assembly,[87] canopy and tail cracks,[88] and brake and fuel vent issues.[89] In 2000, the Air Force and the Secretary of Defense reported projected production cost estimates that exceeded the congressional ceiling of $39.8 billion: $40.8 billion and $48.6 billion, respectively.[90] The 2001 SAR subsequently reported an acquisition program baseline cost breach, leading to a revised acquisition program baseline. The $5.4 billion increase was attributed to decreased aircraft quantity, reduced business base savings because of F-35 procurement, and the loss of potential MYP contracting savings in FYs 2004 and 2005 that had originally been anticipated by the program office. Because of rising costs over its development and production timeline, the Air Force and the prime contractor implemented multiple CRIs. The Productivity Improvement Program (PIP), with a reported $475 million planned investment from 2001 to 2006, aimed at achieving a return on investment of 7:1.[91] Other price reduction efforts during this time included a contractually obligated downward Target Price Curve, where the contractor received financial incentives to achieve the curve;[92] implementation of lean initiatives and "best commercial practices" by LMA, Boeing, and P&W in 2004;[93] and a yearlong tactical Cost Reduction Task Force initiative, later transitioned to the Cost Reduction Organization, that reportedly realized additional savings in lots 5–8.[94]

In an effort to further reduce costs, the Air Force proposed an MYP buy for lots 7, 8, and 9 beginning in FY 2008 after the Defense Appropriations Conference Report directed creation of a report on alternative procurement strategies for the program. The Institute for Defense Analyses provided an initial assessment, completed in May 2006, which estimated savings over annual contracts at $235 million, approximately 2.6 percent for the air vehicle and 2.7 percent for the engine. RAND subsequently evaluated contractor proposed savings and estimated savings of $411 million, or 4.5 percent.[95] Table B.10 shows the characteristics of the program prior to the MYP.

---

[87] DoD, *Selected Acquisition Report: F-22 Raptor Advanced Tactical Fighter Aircraft (F-22)*, December 2001b.

[88] Adam J. Hebert, "Canopy Cracks Ground Raptor Fleet; Replacements Are on the Way," *Inside the Air Force*, May 26, 2000; Elaine M. Grossman, "Air Force Finds Cracks in F-22 Fighter That May Prompt Tail Redesign," *Inside the Air Force*, August 10, 2001.

[89] Laura M. Colarusso, "Air Force Identifies Problems with F-22 Fuel Vent, Brake Proximity," *Inside the Air Force*, January 25, 2001.

[90] GAO, *Defense Acquisitions: Recent F-22 Production Cost Estimates Exceeded Congressional Limitation*, Washington, D.C.: GAO/NSIAD-00-178, August 2000.

[91] DoD, 2001a.

[92] DoD, 2001a.

[93] Defense Acquisition Executive Summary Report, *Assessments for the October 2004 Review*, summary for F-22, PNO 265, October 2004b.

[94] Defense Acquisition Executive Summary Report, *Assessments for the July 2005 Review*, summary for F-22, PNO 265, July 2005c.

[95] Younossi et al., 2007.

Table B.10. Characteristics of F-22 Prior to MYP Award

Characteristics	Contract (Award Year)	Type	Quantity	Cost and Pricing Data Provided
Production contracts	Lot 1 (LMA) (1999)	FFP	10	Unknown
	Lot 2 (1999)	FFP	13	Unknown
	Lot 3 (2001)	FFP	23	Unknown
	Lot 4 (LMA) (2002)	FFP	22	Unknown
	Lot 5 (LMA) (2004)	FFP	48	Unknown
	Lot 6 (LMA)(2005)	FFP	24	Unknown
Concerns about MYP eligibility	• Potential limited savings as later in production and also only a three-year contract			

Award of MYP Contract

Negotiation delays resulted in a number of UCAs because of reported "(1) negotiation delays caused by contractor cost performance difficulties, (2) increased lead time to receive proposals and (3) changing requirements."[96] As final negotiations for the MYP FFP contract concluded, the lot 7 advance buy was awarded on December 23, 2005, followed by the lot 8 advance buy and lot 9 titanium buy a year later (December 21, 2006). Before the final MYP contract award, air vehicle EOQ and the engine EOQ were awarded on January 1, 2007, and February 28, 2007, respectively. Finally, the three-year MYP contracts for 60 aircraft and 120 engine installs were awarded in July 2007.[97] Congress denied the Air Force's request to use incremental funding during the MYP, but it did authorize an unfunded cancellation ceiling. Mirroring RAND's savings estimate, the MYP contract strategy predicted $411 million in savings over annual contracting, or 4.5 percent.[98] RAND analyzed six categories of potential savings: alternative sourcing, production build-out or acceleration, buyout of parts and materials, proposal preparation, support labor, and management challenge. No new CRIs were funded for the MYP because it was at the end of the production run. EOQ funding was provided, however. Of the categories examined by RAND, the largest source of estimated savings by far was buyout of parts and materials, financed by EOQ funding. Also important were support labor and production build-out or acceleration. These factors would also likely be made possible through EOQ funding.

Discussions with the program office after contract execution revealed that LMA achieved almost 10 percent more profit than originally predicted under the FFP contract, leading the program office to speculate that additional savings could have been negotiated before contract

---

[96] Defense Acquisition Executive Summary Report, *Assessments for the October 2006 Review*, Defense Acquisition Executive Summary for F-22, PNO 265, October 2006d. Accessed via the Defense Acquisition Management Information Retrieval database on March 31, 2016.

[97] DoD, *Selected Acquisition Report: F-22 Raptor Advanced Tactical Fighter Aircraft (F-22)*, December 2006b.

[98] U.S. Air Force, *Air Force Signs Multiyear Contract for F-22*, August 8, 2007.

signing.[99] This supports our earlier argument that detailed knowledge of key subcontractors' cost structure is necessary to gain optimal savings. Such an assessment does not appear to have been carried out on the F-22 program, probably in part because of the very short timeline to contract award. Table B.11 lists the basic characteristics of the F-22 MYP.

Table B.11. F-22 MYP I

Characteristic	MYP I
Period of performance	FYs 2008–2010
Contract amount at award (TY$)[a]	$7.4 billion (aircraft), $1.3 billion (engines) (2007)
Last reported contract amount	$7.5 billion (aircraft),[b] $1.4 billion (engines)[c]
Quantities	60 aircraft, 120 engines (later increased to 137)
Contract type	FPEPA/FFP/CPFF/FPIF
CRI funding	$0
EOQ funding	$255 million[d] + $45 million[e] (+ AP for titanium)[f]
Estimated savings	$411 million
Cost and pricing data regularly reported	No

NOTE: CPFF = cost plus fixed fee.
[a] Initial award only included the lot 7 advance buy. The figure listed in the table is the contract price when the subsequent lots and other factors were added.
[b] DoD, *Selected Acquisition Report: FP-EPA Efforts $958.8 Million,* December 2010. FFP efforts: $397.2 million, CPFF efforts: $1.9 million.
[c] "The F-22 Raptor: Program and Events," *Defense Industry Daily*, March 14, 2016. This was an FFP contract modification for "an F-22 multiyear economic order quantity procurement."
[d] "The F-22 Raptor: Program and Events," 2016. This was an FFP contract modification for "F-119 engine multiyear economic order quantity effort, undefinitized contract action."
[e] "The F-22 Raptor: Program and Events," 2016. This included $19.6 million FFP UCA in support of lot 8 aircraft.

Summary and Lessons for F-35 BB

The F-22 MYP contract occurred at the end of procurement, leading to a limited need for comprehensive historical reporting from the contractor for future negotiations, and also greatly reducing the government and contractor's incentives to invest in CRIs. Thus, even though the F-22 offers the most analogous platform comparison to the F-35, the timing of its MYP contract limits the direct lessons learned that could be applied. That said, the profit increase mentioned above underscores the importance for the F-35 program of gaining detailed knowledge of key subcontractors' cost structures to maximize savings.

---

[99] Discussions with representatives from the F-22 program office, December 2015.

MH-60R/S (H-60 MYPs VII and VIII)

Since the 1970s, the Army and Navy have combined their airframe procurements for UH-60 "Black Hawk" and SH-60 "Sea Hawk" helicopters because of their significant design similarities, often using MYP contracts. Both military services' variants have gone through multiple design iterations since their initial versions, and by the late 1990s, the Navy was interested in consolidating its entire helicopter fleet into a limited number of SH-60 variants to lower total operation and support costs.[100] The MH-60R and MH-60S variants were developed for this reason. Initial MH-60R/S production began in the early 2000s, and the first joint procurement of MH-60S and UH-60 helicopters came in a 2002 MYP contract between Sikorsky and the Army, with the Navy involved in contract discussions and in MYP exhibit preparation.[101]

We discuss MH-60R/S MYPs in additional detail because detailed MYP documentation for these contracts is readily available.[102] Including these contracts, the H-60 program has been authorized to execute eight total airframe MYP contracts to date, with the first one being placed in 1982. Throughout the H-60 program, separate MYP contracts have also been placed for the Navy's SH-60 mission system and the aircraft's engine. We focus on the two MYPs (VII and VIII) where the Navy procured both the MH-60R and MH-60S during the same MYPs. We chose these two because of the detailed data readily available: The Navy prepared and submitted separate budget justification documents that included separate estimates of savings and costs for each of the two Navy variants during these two MYPs. During this same period, the Navy also pursued a separate MYP to procure the cockpit avionics and other mission systems for its two variants of the H-60. This was also reported separately in an MYP exhibit in Navy budget documents, making overall examination of the Navy part of the program readily assessable for analysis.

Recent MH-60 MYP (VII and VIII) Contracts

As with the 2002 MYP, the Army was formally designated as the Navy's executive agent for entering MYPs VII and VIII.[103] As the contract's executive agent, the Army was responsible for signing the joint service contract with Sikorsky, but the Navy continued to submit separate MYP budget justifications and obtain separate MYP authority specifically for their H-60 variants.

Contract discussions for MYP VII began once Congress approved multiyear UH-60/MH-60S procurement in 2006 and then approved adding MH-60Rs to the contract in 2007. MYP VII authority resulted in a $7.4 billion, five-year UH-60/MH-60 contract in 2007 that contained options for up to $4.2 billion in additional airframes and spares. The final production quantities

---

[100] This fleet comprised multiple SH-60 variants (B/F/H), HH-1s, UH-3Hs, and UH-46Ds.

[101] H-60 MYPs VII and VIII; see U.S. Senate, *National Defense Authorization Act for Fiscal Year 2001 Report*, 106-292, May 12, 2000.

[102] These MYPs correspond to H-60 program MYPs VII and VIII.

[103] See U.S. Senate, 2000.

for the two variants were not defined by the contracts and were allowed to be changed based on funding allocations, service need, or potential international interest in the program. MYP VIII was a similar five-year, $8.5 billion contract signed in 2012 between the Army and Sikorsky. The contract also contained options for up to $11.7 billion worth of airframes and auxiliary equipment. When combined, the Navy's baseline procurement quantities for MYPs VII and VIII make up approximately a third of the total airframes procured under these contracts.[104] Table B.12 provides summary data for MH-60 MYPs VII and VIII.

While neither contract involves definitive airframe quantities by year, both Army and Navy input is very important to program cost projections. In 2014, changing budgetary priorities raised significant questions about the Navy's expected MH-60 procurement quantities. While some fluctuations in procured airframes could be tolerated, the potential end of the Navy's contractual commitment to MH-60 procurements led to discussions of full MYP VIII cancellation (and the associated cancellation penalty), which has not come to fruition.

Table B.12. Comparison of MH-60 MYP VII and MYP VIII

Characteristic	MYP VII	MYP VIII
Period of performance	FYs 2007–2011	FYs 2012–2016
Contract amount	$3.5 billion	$3.4 billion
Quantities (MH-60R and MH-60S)	260	193
Contract type	FFP	FFP
CRI funding	$0	$0
EOQ funding	$0	$0
Estimated savings	$165 million	$313 million
Cost and pricing data regularly reported	Yes (assumed)	Yes (assumed)

Non-Airframe MYPs

The significant differences in missions for the Army and Navy's H-60 variants are reflected in the different mission/radar systems carried aboard each helicopter variant. Since the beginning of the SH/MH-60 program in the 1970s, the Navy has maintained its own set of contracts for the mission system and cockpit, and formal MYP authority has been provided for recent design iterations of these systems. The mission systems are a combination of Navy-specific avionics and sonar systems that are common to all MH-60R/S helicopters.[105] Recent MH-60R/S avionics systems MYP contracts were signed in 2007 and 2012 between the Navy and LMA, with MYP

---

[104] This is based on 2014 MH-60 SARs procurement quantity projections compared with MYP VII and VIII baselines with no options exercised. DoD, *Selected Acquisition Report: MH-60R Multi-Mission Helicopter (MH-60R)*, Washington, D.C., December 2014b; DoD, *Selected Acquisition Report: MH-60S Fleet Combat Support Helicopter (MH-60S)*, Washington, D.C., December 2014c.

[105] An example is anti-submarine missions.

authority being provided by Congress on the same timeline as MYP authority for the MH-60 airframes. We did not evaluate cost savings associated with MH-60 avionics contracts because contract characteristics for these components were different than contract characteristics for larger contracts, such as those for airframes.

Summary and Lessons for F-35 BB

As discussed in earlier in this appendix, the MH-60 program is a good example of the impact of follow-on multiyear contracts on a program's overall cost savings. Additionally, MH-60 SAR information shows that program savings estimated prior to placement of a multiyear contract can be quite conservative relative to actual savings achieved by the program.

Based on significant differences between MH-60 MYPs VII and VIII and the prospective F-35 BB, lessons from MH-60 contracts may have limited direct applicability to the F-35 program. The contract type (FFP vs. FPIF) and amount of EOQ/CRI funding (none available for MH-60) are just two important examples of differences between the programs. In addition, the MH-60 is a rotary-wing aircraft (not a fixed-wing fighter), and had undergone multiple MYPs prior to the program seeking approval for MYPs VII and VIII. This likely had significant effects on the program's estimated and achieved savings, possibly a key enabler of the achieved level of savings despite no EOQ or CRI funding for these MYPs.

While there have been many H-60 MYPs with Navy and Army collaboration, MYP VII and VIII are the only two so far to include relatively new Navy variants; in this case, the MH-60R and the MH-60S. The Navy published several prenegotiation estimates for the savings for both the second MH-60R/S MYP (H-60 MYP VIII) and the second A/C MYP, as well as actual negotiated settlements post negotiation. While the quantities changed only minimally, the costs and savings percentages reported by the Navy dramatically improved in the final negotiated settlement for both MYPs compared with both the prenegotiation estimates, as well as with the savings estimated for the first MH-60R/S MYP (MYP VII). We did not have the opportunity to interview H-60 program officials or examine program office documentation. However, we speculate that this dramatic improvement may be at least partly because of some of the same factors that affected other programs we examined—i.e., attempts to delve into the lower-supplier tiers to discover true supplier costs and pricing structures as a means to identify greater opportunities for savings during actual contract negotiations.

CH-47F

The Boeing Company's CH-47F Chinook is a multirole, vertical-lift platform that the U.S. Army describes as an essential component of its Army Future Force.[106] Its mission is to transport

---

[106] Elements of this section were taken from previous research by RAND colleagues William Shelton, Stephen Joplin, Cynthia Cook, Abby Doll, James Dryden, Bernard Fox, Mark Lorell, Karishma Mehta, Leslie Payne, Katherine Pfrommer, Soumen Saha, and Cole Sutera. Army Future Force: DoD, *Selected Acquisition Report:*

troops (including air assault), supplies, weapons, and other cargo in general support of operations. The CH-47F enables the Army to support the rapid-response capability for forcible and early entry contingency missions, as well as tactical and operational nonlinear, noncontiguous, simultaneous, or sequential operations, which will be characteristic of future operations.[107] Current U.S. variants in use are the CH-47F and the Special Operations MH-47G. The MH-47G incorporates all of the features of the CH-47F and also has unique, mission-specific equipment.

The Chinook is part of the Army Cargo Helicopter Modernization Program. CH-47Fs are the result of either completely new production units, called new builds, or the refurbishment and conversion of CH-47Ds through a "renewal" process.[108] Additionally, Army Special Forces are receiving MH-47Gs either as new builds or renewed aircraft. All such work is being performed on the same Boeing production line in Ridley Park, Pennsylvania. New engines are produced and old ones refurbished at Honeywell in Phoenix, Arizona, as part of a separate government-furnished equipment contract.

Characteristics Prior to MYP

The CH-47F acquisition strategy was developed initially for the Improved Cargo Helicopter Modification Service Life Extension Program. This program extended the life of aging CH-47Ds to fill the gap until a new design helicopter was developed and procured to replace the CH-47. Two years of LRIP of 30 aircraft had been authorized at Milestone II in December 1997.[109] However, in 2003 and 2004, the program was restructured in part because of expanding requirements stemming from the terrorist attacks of September 11, 2001. The CH-47F program temporarily stood down in 2003 so Boeing could focus on remanufacturing CH-47Ds into MH-47Gs for Special Operations units (under a separate acquisition program). MH-47G production received higher priority from OSD in part because of the war in Afghanistan.[110]

As a result of this prioritization, the program office submitted a program deviation report to the Under Secretary of Defense for Acquisition, Technology, and Logistics warning of the expected schedule slip or production hiatus for the CH-47F.[111] During this CH-47F production hiatus, engineers expanded the upgrades planned for the CH-47D to CH-47F remanufacture to

---

CH-47F Improved Cargo Helicopter, December 2014a. The U.S. Army is executing an ongoing future force initiative that allows it to address the challenges of the 2025 time frame. The CH-47F and other aircraft platforms are components of the Army Future Force. More information on the Army Future Force is found in Brian Nichiporuk, *Alternative Futures and Army Force Planning*, Santa Monica, Calif.: RAND Corporation, MG-219-A, 2005.

[107] DoD, 2014a.

[108] The term *renew* or *renewed* is used to describe CH-47 aircraft that are being rebuilt and upgraded. The term is found in various CH-47F SARs and related PBs and is common nomenclature used by the CH-47F program management office in Huntsville, Alabama.

[109] DoD, *Selected Acquisition Report: CH-47F Improved Cargo Helicopter*, December 2007.

[110] DoD, 2007; DoD, 2003.

[111] DoD Office of the Inspector General, 2004.

meet more-demanding capability requirements. In addition, the decision was made to produce entirely new CH-47Fs simultaneously. The program was approved for production in November 2004 with a revised acquisition strategy that included the following features:

- concurrent production of both new and remanufactured CH-47Fs
- a focus on manufacturing and operating cost-reduction improvements, particularly through the use of MYP fixed-price contracts
- more-extensive airframe and system capability upgrades for new and renew aircraft, to be implemented first in lot 3 (FY 2005)
- development of a preplanned, phased upgrade program[112]
- CPFF contracts were retained for research, development, test, and evaluation and procurement of developmental prototypes.

Incentives were provided for the initial annual LRIP lots with an FPIF contract for refurbished CH-47F, and a FFP contract for new-build CH-47F. Both contracts had annual options. Boeing began a large capital investment in its Chinook manufacturing lines, with two new lines, one for the F model and another for all other Chinook models, completed in 2011.[113] The revised CH-47F acquisition strategy also focused heavily on cost reduction measures and design stability and, as a result, moved toward a strategy of MYP contracts for future procurements, each of which included upgrades the Army hoped would be incorporated.[114] Table B.13 provides a summary of the CH-47 contracts executed prior to the transition to MYP contracts.

---

[112] DoD, *Selected Acquisition Report: CH-47F Improved Cargo Helicopter*, 2004.

[113] Bill Carey, "Boeing Renovates CH-47 Chinook Line for Increased Production," AINonline, July 1, 2011.

[114] Not all of the upgrades were fully funded or contracted for at the onset of both MYPs. With each airframe, the refurbishment became more extensive and required more replacement parts as the program progressed. Regarding major equipment modification, we do not know when specific modifications cut into the production line. However, we do know that much of the MYP I equipment modifications were conducted at the depot rather than on the production line. In MYP II, the plan was to move the modifications to the production line. Similar to the MYP I modifications, we do not have information on when these cut-ins occurred.

Table B.13. Characteristics of CH-47F Prior to MYP Award

Contract (Award Year)	Type	Quantity	Cost and Pricing Data Provided
LRIP lot 1 "F" (2002)	FPIF	1	Unknown
LRIP lot 1 "G" (2002)	CPFF	6	No
LRIP lot 2 (2003)	CPIF	16	No
New build recurring (2004)	FFP	59	Unknown
FRP lot 3 "F" (2005)	FPIF	8	Unknown
FRP lot 4 "F" (2005)	FPIF	15	Unknown
FRP lot 3 and 4 "G" (2005)	FPIF	18	Unknown
FRP lot 5 "F" (2005)	FFP	9	Unknown
FRP lot 5 "G" (2006)	FFP	6	Unknown
FRP lot 6 "G" (2007)	FFP	6	Unknown
FRP lot 7 "G" (2008)	FFP	6	Unknown

Award of First MYP Contract

In 2008, the first five-year MYP contract was signed for a total of 215 CH-47F aircraft. In total, the contract included 109 new-build aircraft, 72 renew aircraft, and priced options for 34 new-build aircraft. By the time of contract award, the funded first year was already exercising an option for ten new-build aircraft with FY 2008 supplemental funding. In December 2008, the second year (lot 7) was partially obligated for 16 new-build and 15 renew aircraft, followed in April 2009 by completed obligation of lot 7, with seven additional new-build aircraft.[115]

Differing interpretations of FAR Part 15 delayed negotiations and limited time for the government to fully assess and evaluate Boeing's cost and pricing data. The disagreement lay in when certified subcontractor cost and pricing data must be supplied during negotiations. Boeing asserted that this information could be supplied at any time and that historical data and initial quotes were adequate initially, while DCMA argued that the prime contractor should submit the data before or at the time of proposal submission. Because of the delayed submission of final data, Boeing held a large informational advantage over the government during negotiations. After it negotiated the final, fixed-price contract, Boeing was able to raise profits through tougher negotiations with its suppliers, and savings that were not passed on to the government. The government, however, took note and applied these lessons learned to the next MYP negotiations.[116]

---

[115] DoD, *Selected Acquisitions Report: CH-47F Improved Cargo Helicopter (CH-47F)*, December 2009.

[116] Discussions with Boeing DCMA, March 2015.

Award of Second MYP Contract

MYP II was signed in 2013 for a total of 155 renew and new-build CH-47Fs, with an option for 60 additional aircraft, of which 23 were exercised, resulting in a revised total of 178 CH-47F aircraft. Initially, in December 2010, the Army program office estimated potential savings of $373 million, or about 10 percent, over the use of annual contracts. By the time the contract was awarded in June 2013, however, the predicted savings had risen to $810 million, or about 19 percent.

This dramatic increase was due to a number of factors, primarily drawn from lessons learned from the previous negotiations. The government required Boeing to include actual negotiated cost and pricing data from subcontractor negotiated agreements, not just historical data or initial quotes. If the supplier competed directly with the prime in a specific area and did not want to provide detailed proprietary cost data to the prime contractor, the government directly received the data instead of going through the prime. Subcontractor proposal quality was also improved through the use of a pricing template, instructional videos, and direct government oversight and engagement with suppliers on proposal standards. Although the time from RFP to proposal submission may have taken longer, the overall time from RFP to contract award was not as affected, and the government was able to obtain a better deal through increased visibility. DCMA stated that to motivate Boeing to cooperate, it used a "carrot" of two additional MYP contracts and a "stick" of auditing business systems for compliance if supplier proposals continued to be of poor quality.[117] Finally, a CAPE analysis concluded that annual contracts for 155 CH-47Fs would cost 27 percent, or $4.2 billion, more than the Army's estimate. This, in addition to adjusted negotiation practices, led to an almost doubling of predicted savings for the MYP.[118]

Despite the savings from the MYP contract, an Inspector General report asserted that Boeing significantly overcharged for safety stock costs (around $36.8 million) and eight parts ($10.6 million to $19.1 million). Boeing, however, adjusted the requirements after review by U.S. Army Aviation and Missile Command.[119]

---

[117] Discussions with DCMA, March 2015.

[118] Jason Sherman, "Higher CAPE Estimate Amplifies Army Savings Claim in New CH-47F Contract," *Inside the Pentagon*, July 4, 2013.

[119] Jen Judson, "IG: Boeing Overcharged for CHCH-47F Parts and Inefficiently Used Funds," *Inside the Army*, July 22, 2013.

CH-47F is a good example of the importance of extensive preparations by the government prior to negotiations for maximizing cost savings on an MYP contract, the results of which are demonstrated in Table B.14—specifically, the increase in savings between the original estimate and as awarded contracts. Based on a thorough assessment of actuals near the end of the first MYP, CH-47F program office officials concluded that many savings opportunities for the government had been overlooked or missed on the first MYP for two major reasons. One was a disagreement over the point in negotiations at which the prime contractor must supply TINA-compliant data on its subcontractors' pricing and cost data. This is an issue of dispute that is not resolved in the FAR Part 15 regulations. Historically, many prime contractors have provided the government initial quotes from their subcontractors in the initial contractor response to the RFP, which are used as the basis for negotiation. Once the prime contract is negotiated, or nearly negotiated, the prime can go back and negotiate much tougher deals with its subcontractors, thus enhancing its profit margins. The CH-47F program office believed that this happened in the case of the MYP I, and developed strategies for the second MYP that required final negotiated cost and pricing data between the prime and subcontractors in time for the government to use these data in negotiating the prime contract. Second, the government realized that the lower-tier contractors escape close government scrutiny and thus have considerable freedom on pricing without the government being fully aware of their true cost structures. The program office decided to devote more time and resources to determining the true overall price structure of key subcontractors prior to the negotiation of the prime contract. In this way, the government was much more aware of where cost savings might be available and where it could press hard for savings in negotiations.

Table B.14. CH-47F MYP I and MYP II

Characteristic	MYP I	MYP II
Period of performance	2008–2013	2013–2018
Contract amount at award (TY$)	$2.4 billion (2008)[a]	$1.5 billion (2013)[b]
Last reported contract amount (TY$)	$4.4 billion (2013)	$2.8 billion (2015)
Quantities	116 (later increased to 215)	65 (later increased to 105)
Contract type	FFP	FFP
CRI funding	~$1 million	$0
EOQ funding	$0	$0
Estimated savings	$450 million, 10.76%[c]	Original estimate: $373 million, 10% At award: $810 million, 19%
Cost and pricing data regularly reported	No	Yes

[a] The initial contract price was $722.7 million for 35 aircraft. Modifications and exercised options, however, led to an increase in both price and quantity as shown. DoD, 2009.
[b] The initial contract price was $916.5 million for 37 aircraft, but the contract was soon increased to the quantities and price shown in the table. See U.S. Air Force, 2014a.
[c] DoD, *NDAA Sectional Analysis*, March 12, 2012.

This program and several others we examined make it clear that the actual negotiating process may be far more critical for actual MYP savings than the estimating and justification process used to gain government approval. The estimated costs and savings that are presented to Congress rarely play a central role in the actual contract negotiations, according to our sources. Furthermore, there is never any attempt to validate whether the savings used in the budget justifications were ever achieved—although the actual negotiated contract and the savings obtained are more easily verified and validated.

V-22

The V-22 Osprey is a unique hybrid of a vertical takeoff/landing rotary-wing aircraft and a fixed-wing turboprop aircraft. Development of this hybrid "tilt-rotor" technology began in the early 1970s to combine the over-land speed of a conventional aircraft with the takeoff/landing flexibility of a rotary-wing aircraft.[120] This combination of capabilities was desirable because it had the potential to satisfy the operational needs of multiple military services.[121] While early tilt-rotor aircraft developments initially were heavily Army funded, the V-22 program shifted to a four-service, Navy-led joint program by the early 1980s. The first V-22 development contract was awarded in 1986 and was an FPIF contract jointly awarded to Bell Helicopter Textron and Boeing-Vertol.

Characteristics Prior to MYP

Significant technical issues, including fatalities during crashes of prototype V-22s, led to a major program restructuring in 1992. This restructuring included cancellation of the early production contracts, the Army withdrawing from the program, and major design modifications to improve aircraft safety. The significance of the design modifications led to the cancellation of the existing development contract and the awarding of two new contracts: one for the engine, and one for a modified airframe to meet the Marine Corps requirements (later known as the MV-22). Under this contract, LRIP was scheduled to begin in 1996 and FRP would begin in 2001. To accommodate slight differences in operational need between Marine Corps and Special Operations missions, a second carrier variant (the CV-22) was added to the program in 1995. The two variants have similar airframes, but the MV-22 was designed with personnel support missions in mind while the CV-22 was specially designed with capacity for low-altitude and long-range special operations.

---

[120] *Tilt-rotor* is defined as the ability to rotate the aircraft's wing-mounted engines from a forward-facing arrangement for overland flight to a vertical arrangement for vertical takeoff/landing.

[121] Specifically, these were operational compatibility with short-deck amphibious assault ships for Marine Corps, combat search and rescue capabilities for the Navy, and ability to perform Special Operations missions of Air Force/Army/Special Operations Command.

The first two LRIP aircraft lots were delivered in 1999 and 2000, but delays resulting from additional technical issues caused further delays to the 2001 FRP decision. OSD appointed an independent panel (the "Blue Ribbon Panel") to evaluate the overall performance of the program and determine when it would be ready to enter into FRP. To maximize cost savings, the Blue Ribbon Panel determined that MYP contracts should be pursued for the V-22 once it had entered FRP. By 2003, design changes had been implemented and prototype tests were proceeding satisfactorily, so the program office ramped up the V-22 production rate. The V-22 was authorized to enter FRP in 2005. A summary of all V-22 pre-MYP contracts can be found below in Table B.15.

Table B.15. Characteristics of V-22 Program Prior to MYP Award

Contract (Award Year)	Type	Quantity	Cost and Pricing Data Provided
LRIP 1 (1997)	CPIF	5	Yes (assumed)
LRIP 2 (1998)	CPIF	7	Yes (assumed)
LRIP 3 (1999)	CPIF	7	Yes (assumed)
LRIP 4 (2000)	FPIF	11	Yes (assumed)
LRIP 5 (2001)	FPIF/CPIF/CPFF	9	Yes (assumed)
LRIP 6 (2002)	FPIF/CPIF/CPFF	11	Yes (assumed)
LRIP 7 (2003)	—[a]	11	Yes (assumed)
LRIP 8 (2004)	FPIF/CPIF	11	Yes (assumed)
LRIP 9 (2005)	FFP/FPIF	11	Yes (assumed)
FRP 10 (2006)	FFP	11	Yes (assumed)
FRP 11 (2007)	FFP	20	Yes (assumed)
Concerns about MYP eligibility	• Political issues surrounding program because of technical problems and crashes early in program life		

[a] Contract type for LRIP 7 was not listed in applicable V-22 SARs and other contract documents.

Award of First MYP Contract

By the time that the first MYP was approved for the V-22 program in 2008, 114 aircraft had been procured as part of 11 annual contracts. The first MYP was valued at $10.4 billion for 175 aircraft, but did not include engines, which were contracted for separately.[122] The estimated total cost savings for this contract was a relatively modest 4 percent, credited largely to a combination of EOQ and government-funded CRIs. These savings are discussed in detail in the V-22 program's MYP justification exhibits.[123] Other sources of cost savings identified by the

[122] According to V-22 SARs, the V-22 engine has been separately contracted for with RR through multiple separate non-MYP contracts over the course of the V-22 program.

[123] The MYP exhibits were submitted in early 2006 to support a 2008 contract award.

MYP exhibits include reduced administrative costs for the government and contractors (proposal, negotiation, purchase order, and contract management), improved production planning, long-term agreements with first- and lower-tier suppliers, and additional use of competition in component procurement. To provide additional incentives for good cost performance from the suppliers, the government provided generous share-line distributions for cost underruns, while maintaining a higher level of risk for cost overruns. A separate share line was even established for the parts of the contract with the highest supplier cost risk because of market volatility in specialty metals.[124]

The V-22's first MYP exhibit also discusses how program stability and the presence of a realistic program cost estimate were proven to Congress. Aircraft configuration stability was largely demonstrated through the test and training hours undergone since the late 1990s and early 2000s. Few additional major technical changes were planned for the V-22 once FRP had begun, so the configuration that had been demonstrated to date would be very close to the MYP I configuration. To ensure the program had a realistic cost estimate, the V-22 program office worked with independent estimating teams from the Naval Cost Analysis Directorate, the Air Force Cost Analysis Agency, and OSD Cost Analysis Improvement Group. The lowest of these independent estimates became the final requested program funding level in the MYP exhibits. Despite proven cost realism and a stable aircraft configuration, the program's negative publicity to date made getting congressional approval for the MYP difficult.[125] Ultimately, Congress authorized the MYP to promote additional stability in the program and keep aircraft unit costs low.

Award of Second MYP Contract

Planning for the next MYP continued during execution of the first MYP, and the contract was awarded in June 2013. According to 2012 V-22 MYP exhibits, the contract included the procurement of 98 aircraft for $6.5 billion and was estimated to save 11.6 percent when compared with annual contracting. The key sources of savings cited by the MYP exhibits are similar to the savings sources cited for other MYP contracts: EOQ (though less was authorized than for the previous MYP contract), long-term supplier agreements, improved production planning, reduced obsolescence risk through parts' buyouts, alternate and/or competitive sourcing, reduced administrative costs for the government and contractors, and deep-dive negotiation tactics with suppliers (including looking into savings between higher- and lower-tier suppliers). No government funding for CRIs was included with the MYP II contract, although the Bell-Boeing team pursued its own CRIs to improve efficiency further.

---

[124] The two sets of share line were 70:30 and 20:80 (government:industry) for cost overruns and underruns, respectively, and 90:10 and 20:80 for specialty metals.

[125] Discussions with the V-22 program office personnel, December 2015.

The fact that the V-22 had MYP cost data from the previous five years greatly simplified demonstrating program stability and cost realism, and was a key factor in receiving approval for the second MYP. The greater savings for the second MYP was attributed to three major factors: increased MYP experience, deep-dive negotiation tactics, and the presence of OSD's Better Buying Power Initiatives.[126]

Discussions are under way for a third MYP contract to begin in 2018, immediately following the end of the second MYP. The Naval Air Systems Command (NAVAIR) is considering opening the procurement of V-22s to international allies to increase the number of aircraft procured on the contract. NAVAIR is also considering creating another V-22 variant to replace the Navy's aging C-2 Greyhound platform and including this variant in the third V-22 MYP.[127] The third MYP will involve fewer total aircraft than the previous two, as the V-22 force levels are reaching their planned maximum levels.

Summary and Lessons for F-35 BB

The V-22 program shares one notable characteristic with the prospective F-35 BB: The program did not enter its first MYP until after multiple years of LRIP annual contracting. Despite this similarity, many program attributes displayed in Table B.16, including the total contract amount and number/type of procured aircraft, are significantly different for the two programs. Another key programmatic difference is the level of EOQ and CRI funding available to the V-22 program. As shown in Table B.16, less than 0.5 percent EOQ funding was available for V-22 MYP I and EOQ funding for MYP II was limited to approximately 2.5 percent. Though V-22 MYP I was for a similar total value to the prospective F-35 BB, about half of the planned CRI funding for the F-35 BB was available for V-22 MYP I. No CRI funding was provided for MYP II.

Table B.16. Comparison of V-22 MYP I and MYP II

Characteristic	MYP I	MYP II
Period of performance	FYs 2008–2012	FYs 2013–2017
Contract amount	$10.4 billion	$6.5 billion
Quantities (MV-22 + CV-22)	175	98
Contract type	FPIF	FPIF
CRI funding	$170 million	$0
EOQ funding	$33 million	$160 million
Estimated savings	$427 million	$852.4 million
Cost and pricing data regularly reported	Yes (assumed)	Yes (assumed)

---

[126] Discussions with the V-22 program office personnel, December 2015.

[127] This is a carrier onboard delivery platform that performs logistics/resupply missions to carriers at sea.

Despite these significant programmatic differences, the first V-22 MYP contract was estimated to be able to achieve a similar percentage savings to what RAND projects the F-35 BB could achieve, or about 4 percent overall. V-22 MYP II achieved an even higher percentage savings than MYP I, more than 10 percent of the total contract value. This suggests that follow-on multiyear lessons benefit from learning, especially when cost data from the initial MYP shows differences in the expected cost and profit rates. Like the CH-47F program, the V-22 program office was determined to negotiate a much better deal for the government for MYP II compared with MYP I based on more-extensive knowledge of the actual costs of the V-22 supplier base. This is an important lesson from the V-22 program for the future potential savings for F-35 BBs and MYPs.

Virginia-Class Submarine

The *Virginia*-class submarine design emerged in the early 1990s as a replacement for two existing classes of attack submarine, the workhorse *Los Angeles* class and the revolutionary *Seawolf* class. Both had been designed during the Cold War, and while the extremely capable *Seawolf*-class design had only just been completed, the early *Los Angeles*-class submarines were aging and would require either an expensive nuclear refueling or replacement by the mid-2000s. An affordable alternative to the *Seawolf* class was required; at the end of the Cold War, funding was not available for an expensive platform with capabilities that were beyond what was deemed necessary for the post–Cold War threat climate.

Characteristics Prior to BB and MYP

As the *Virginia*-class design matured through the mid-1990s, the Navy began discussions with DoD and Congress regarding construction and contract arrangements. The Navy wanted to begin *Virginia*-class procurement in 1998, construct two *Virginia* submarines a year after completion of the first four in 2002, and to sole-source the design and construction contract with no competition. Competition for previous submarine programs existed between General Dynamics–Electric Boat (EB) and the other major nuclear vessel construction shipyard, Newport News Shipbuilding. Because of funding limitations and congressional opposition to constructing submarines noncompetitively, the 1996 National Defense Authorization Act (NDAA) required the Navy to alternate ship construction between the two shipyards for the first four boats and allow competition for the construction contract after that. The 1996 NDAA approved competitive procurement of *Virginia*-class submarines on an annual basis between 1998 and 2002, along with manufacturing one submarine a year for the first four submarines.

In response, the Navy proposed a new cooperative submarine construction approach that combined new manufacturing practices with a five-year balanced workload involving both shipyards that achieved cost savings through long-term guarantees of business.[128] The long-term guarantee also alleviated contractor concerns with the cooperative arrangement; namely, that the shipyards would be required to share proprietary information during construction, which could affect their future competitive contracts. As a result of these benefits, the Navy was very interested in obtaining some sort of multiyear contract authority despite the fact that the *Virginia* class was seen as unqualified for an MYP because of the lack of a mature, stable, proven design.

Award of BB Contract

In the proposed arrangement, EB would remain the sole design agent for the *Virginia* class and would be the construction prime contractor, with Newport News serving as a major subcontractor responsible for approximately 50 percent of the construction effort. The Navy estimated a savings of approximately $640 million for the four submarines based on a multiyear contract estimate from EB. Congress agreed to this approach, later called a "block buy," based on its consideration of cost savings and maintaining steady business at both contractors.[129] congressional agreement included approval of AP funds for the second, third, and fourth submarines; no establishment of a cancellation penalty; and use of a combination of CPFF and CPIF contracts for the first four submarines.[130]

The duration of the BB for *Virginia*-class Block I (SSNs 774–777) was from 1998 to 2002, at which point the Navy had originally planned to increase production levels to two boats per year rather than one. Owing to funding constraints and projected cost growth for the whole class identified in the early 2000s, Congress would not approve this production quantity increase and as a result, future block contracts were not immediately approved following the Block I contract.[131] To keep *Virginia*-class production active as program cost reductions were being pursued, a single submarine was procured in 2003 (SSN 778). For the 2004 NDAA, the Senate agreed to a FPIF MYP contract for five submarines over five years (SSNs 779–783; Block II).[132]

Award of MYP Contracts

An MYP (rather than BB) contract was pursued for Block II because formal government commitment to an MYP contract increased supplier willingness to pursue internal production process investments to further improve *Virginia*-class program savings. EOQ funding was provided for Block II as well as later MYP contracts. EB was given the authority to allocate the EOQ funding to subtier suppliers as necessary to achieve maximum cost reductions.

---

[128] These manufacturing practices involved constructing the submarine in ten separate cylindrical modules that would be individually populated and welded together at alternating shipyards.

[129] U.S. Senate, *National Defense Authorization Act for Fiscal Year 1998 Report*, Report 105-29, June 17, 1997.

[130] For the SSN 774 and 777, CPFF was used; for SSN 775-776, CPIF.

[131] This refers to submarine configuration blocks, not BB contracts.

[132] U.S. Senate, *National Defense Authorization Act for Fiscal Year 2003 Report*, Report 107-151, May 15, 2002.

An additional source of savings on the Block II MYP contract was the use of capital investment incentives (CAPEX) to help shipyards pursue manufacturing/construction cost savings through facility upgrades. The Navy agreed to fund part of the facility improvement cost before construction began and the remaining portion after savings had been realized in future years.[133] This strategy was also used to great effect in follow-on MYP contracts and is a popular Navy procurement case study because of its success in incentivizing construction savings.

The Navy still desired to manufacture *Virginia*-class submarines at a rate of two per year to ensure long-term submarine force levels met their shipbuilding goals. This was made challenging because Congress had cut *Virginia*-class program funding through 2011. The acquisition cost of each submarine now had to be reduced by almost 20 percent to enable a construction rate increase. The resulting cost-reduction effort focused on design changes and measures to increase shipyard efficiency that ultimately lowered per-submarine procurement costs from approximately $2.4 billion to $2 billion (FY 2010 $).[134] The key features of the cost-reduction program included design changes targeted at lowering large recurring costs and MYP contracting with EOQ authorization. With these changes, the five-year Block III (MYP II) FPIF contract was placed in 2009 for eight submarines, with procurement of two submarines per year beginning in 2011. The success of the cost-reduction effort was largely the result of the close cooperation between the Navy program office and the shipyards, which were also incentivized by the contract and CAPEX program to meet the new cost goal to ensure the increased profits associated with manufacturing of two submarines per year.

Following the Block III (MYP II) contract, an FPIF contract for ten Block IV submarines was placed in 2014 through 2018. While cost-reduction investments were heavily pursued for previous blocks, major savings for Block IV were pursued through involved contract negotiations with EB and their subtier suppliers, which caused the negotiation process to take almost twice as long as for Block III, and involved hiring additional program office personnel and consultants. The government's aggressive cost positions during these negotiations were based on detailed cost-performance reporting data that had been obtained from previous MYPs, specifically costs and profits seen between EB and their second- and third-tier suppliers.[135] A summary of all *Virginia*-class program awarded BB and MYP contracts can be found in Table B.17.

A future *Virginia*-class MYP (Block V) is under discussion, but many details of these discussions are not publicly available, other than the fact that more-significant design changes

---

[133] The contract permitted the Navy to recover up to all its investment in specific CAPEX measures if savings could not be demonstrated.

[134] This came to be known as the "design for affordability" program. Design for affordability was pursued using dedicated government investment to achieve savings beyond what would be pursued by EB through the contract share line.

[135] Specific savings values are considered business sensitive and were unavailable.

are planned for Block V than for previous blocks to accommodate the insertion of additional vertical missile launch capability to the front third of the ship.

Table B.17. *Virginia*-Class Submarine Program Summary

Block	Contract Time Frame	Contract Type	Number Procured and Production Rate	EOQ Approved?
Block I (SSNs 774–777)	FYs 1998–2002	BB 774 and 777 CPFF 775 and 776 CPIF	4; 1 per year	No, but AP funds for 775–777
Block II (SSNs 779–783)	FYs 2004–2008	MYP I 779–783 FPIF	5; 1 per year	Yes
Block III (SSNs 784–791)	FYs 2009–2013	MYP II 784–791 FPIF	8; 1 per year 2009-2011, 2 per year 2011–2013	Yes
Block IV (SSNs 792–801)	FYs 2014–2018	MYP III 792–801 FPIF	10; 2 per year	Yes

Summary and Lessons for F-35 BB

While the design complexity and unit cost of a submarine are different from an F-35,[136] applicable BB and MYP contract execution lessons can be learned from the *Virginia*-class program. First, while proving sizable cost savings is beneficial to obtaining BB authorization from Congress, demonstrating other benefits (e.g., maintaining industrial base health) can also be a factor. Second, obtaining EOQ and CRI funding is a significant source of savings for multiyear contracts, and providing the prime contractor the opportunity to direct this funding to specific focus areas is one method to achieve CRI savings. CAPEX is an example of this. For the capital investments developed with CAPEX funds, EB had an opportunity to recommend specific initiatives that it felt would be most effective at reducing production costs and were able to achieve significant savings through these initiatives.

In discussions with multiple program offices that used government CRIs to achieve savings, it was determined that using profit to initiate a project is not necessarily in the financial interest of the supplier for many capital investment projects. This is because the cost of many capital improvement projects is so high that the supplier will not recoup program savings for many years after the initial investment is made. Government investment in CRIs demonstrates long-term commitment or investment to a program, which can help motivate suppliers' capital improvements plans. Another benefit is that it enables suppliers to keep more of their profit in the

---

[136] The average *Virginia*-class submarine unit cost is typically quoted as being between \$2 billion and \$2.5 billion, depending on which block and fiscal year's inflation are being referenced.

short term (rather than having to reinvest in capital improvement projects), which can be used by the government as leverage in contract negotiations.[137]

Lastly, learning lessons from previous contracts within a program is very important to improving savings. For each *Virginia*-class multiyear contract, a different savings mechanism played a significant role in program savings: for Block II, shipyard capital investments; for Block III, design changes to reduce production costs; and for Block IV, deep-dive negotiations with contractors. While some of these apply to annual as well as multiyear contracts, certain savings sources are only achievable or improved by multiyear contracts (e.g., EOQ funding, long-term guarantees of business), and having the benefit of previous multiyear contract experience is beneficial to achieving program savings. In each case, detailed cost reporting information, such as Cost and Software Data Reporting data, was used to evaluate actual supplier costs compared with each placed multiyear contract. As discussed, this evaluation improved the government's position during negotiations for future multiyear contracts. Programs in various stages of execution may find different sources of savings via cost reporting data from previous contracts, but multiple program offices that we interviewed cited learning from previous contract cost data as a key source of savings for their programs.

LCS

The LCS program began in 2001 with the objective of designing a small, fast, inexpensive surface ship that could operate in shallow waters closer to the shoreline than larger naval combatants. LCSs were also designed to carry a single modular "mission module" that can support one mission at a time to reduce platform costs, and that is interchangeable with other mission modules. Mission modules planned for LCS include antisubmarine warfare, Special Operations support, and mine countermeasures. With these mission modules, the LCS is expected to replace multiple aging ship classes depending on configuration. Having a large number of reconfigurable ships in the fleet is seen as an inexpensive way to maintain a high worldwide fleet presence to deter naval conflict.[138]

Contracts for LCS design and construction were placed in 2004 with Lockheed Martin and General Dynamics. Each shipyard was contracted to design and build one ship with an option for a second—for a total of either two or four ships, depending on whether the option was exercised. The two designs that resulted from this competition were very different hull forms with unique onboard control systems and mission modules. The Navy was authorized to exercise the option to purchase a second ship of each type in 2009, resulting in the commissioning of two of each LCS variant by 2014. Also in 2009, the LCS program decided that, rather than continuing to

---

[137] This sentiment was based on anecdotal evidence and not detailed documentation of savings, but was expressed by multiple program offices that we interviewed.

[138] U.S. House of Representatives, *Approving Purchases of Littoral Combat Ships*, Vol. 156, No. 166, December 15, 2010.

build two LCS variants for future contracts, a single design would be selected and ten of the selected design would be procured between 2010 and 2014. After this, the construction contract would be open to competition, with the winning team producing five ships of the previously selected design. This approach was intended to use competition to drive ship acquisition cost as low as possible. Lowering acquisition cost for the LCS program was important because of design phase cost growth and Congress's desire to recover these losses through reduced procurement costs.

Having received favorable proposals in late 2010 from each supplier for FPIF contracts, the Navy petitioned Congress to accept both bids and procure ten of each LCS variant between 2010 and 2015 rather than down-selecting to a single design. The benefits of this approach were the elimination of the "tool-up time" related to potentially having one shipyard build a ship designed by another shipyard, the ability to procure 20 LCS platforms for less than was being carried in the budget for 19, and maintaining long-term business at multiple shipyards. Negative aspects of this approach were related to lifecycle costs of maintaining two ship designs, including training, maintenance/upkeep, and mission module design.[139] Congress agreed to the Navy's recommendation and the contracts were finalized in December 2010.

Award of BB Contract

The LCS contracts were approved as BB contracts rather than MYP contracts for two main reasons. First, by the time that multiyear contracting authority for LCS was being discussed, obtaining the required CAPE cost validation before the drafting of MYP legislative language was not possible by the expiration date of the bidders' proposals. Second, Congress would not agree to a cancellation fee for the LCS contracts so that the procurement of a particular year's vessels could be canceled without penalty. A cancellation fee was undesirable to Congress because of the program's previous cost fluctuations and limited political support for the program. EOQ purchasing was also not authorized for the LCS BB contracts. This was because separate bids with and without EOQ were requested of the suppliers, which identified minimal EOQ-related savings.[140] The main source of savings identified in these bids was the long-term guarantee of business, although based on discussions with the LCS program office, the competitive arrangement likely generated the most savings of any savings category. The LCS BB contracts are summarized in Table B.18.

---

[139] Douglas Elmendorf, informational letter regarding Navy Littoral Combat Ship procurement plans, Washington D.C., December 10, 2010

[140] Discussion with LCS program office, 2015.

Table B.18. LCS Program Summary

Block	Contract Time Frame	Contract Type	Number Procured & Production Rate	EOQ Approved?
Block Contracts 1 and 2	FYs 2010–2015	2 BB, both FPIF	20; 2 contracts for 1 ship per year for 2010–2011 and 2 per year for 2012–2015	No, but AP funds

The Navy planned to purchase 52 LCSs over the life of the program, but over time, planned design concept and mission emphasis have changed significantly. In 2014, the Secretary of Defense ordered the final 20 LCSs (ships 33–52) to be designed as more traditional Naval combatants.[141] This plan was modified in December 2016 when Secretary of Defense Ashton Carter told the Navy to reduce planned LCS procurement from 52 to 40 ships. It is unclear how this will affect future LCS multiyear contracts or the configuration of future ships of the class.

Summary and Lessons for F-35 BB

As with the *Virginia*-class program, LCS BB experience also contains lessons for the F-35 program. Similar to the *Virginia*-class program, long-term guarantees of business for the specialized industrial base and deep-dive negotiations into lower-tier supplier contracts were significant sources of savings for LCS. Another similarity between these two shipbuilding programs was the fact that formal single and multiyear proposals were not obtained from suppliers before pursuing multiyear contract authorization; each program had a multiyear proposal and developed its own annual contracting estimate to compare it in advance of discussions with Congress. Interestingly, the effect of EOQ and CRIs for LCS differed from the *Virginia*-class program; no EOQ was authorized for LCS and no government-funded CRIs were pursued. As discussed, it is likely that competitively developed proposals overshadowed both of these sources of savings.

Possibly the most significant difference between the LCS and F-35 contracts that must be considered when evaluating applicable LCS lessons learned is the fact that the LCS BB contract proposals were developed in a competitive environment. Based on discussions with multiple program offices that have multiyear contract and competitive down-selection program experience, competition for a sole-source winner-take-all contract is likely the most effective source of savings for any program and should be utilized to lower costs whenever possible. While the F-35 is a sole-source contract, adding competition to contracts for lower-tier suppliers could be a significant source of savings for the program. While competitive contract savings are not exclusive to multiyear contracts, the combination of multiyear contracts (and associated government commitment to long-term business) with competition can assist prime contractors in achieving the best price possible from their lower-tier suppliers.

---

[141] Design preferences include, for example, battle hardness at the expense of speed and moving from a reconfigurable ship concept to a traditional platform.

References

"Amid Criticism, Air Force Says C-130J Multiyear Funding Profile Is Sound," *Inside the Air Force*, June 21, 2002.

Bakhshi, V. Sagar, and Arthur J. Mandler, *Multiyear Cost Modeling*, Fort Lee, Va.: Army Procurement Research Office, Office of the Deputy Chief of Staff for Logistics, APRO 84-03, February 1985.

"C-130J Upgrade Targets Aging Transport Fleets," *Aviation Week & Space Technology*, Vol. 139, No. 2, July 12, 1993.

Carey, Bill, "Boeing Renovates CH-47 Chinook Line for Increased Production," AINonline, July 1, 2011. As of August 18, 2017: http://www.ainonline.com/aviation-news/defense/2011-07-01/ boeing-renovates-ch-47-chinook-line-increased-production

Colarusso, Laura M., "Air Force Identifies Problems with F-22 Fuel Vent, Brake Proximity," *Inside the Air Force*, January 25, 2001.

DCMA—*See* Defense Contract Management Agency.

Defense Acquisition Executive Summary Report, *Assessments for the May 2003 Review*, summary for C-130J, PNO 220, May 2003. Accessed via the Defense Acquisition Management Information Retrieval database on March 31, 2016.

———, *Assessments for the January 2004 Review*, summary for F/A-18E/F, PNO 549, January 2004a. Accessed via the Defense Acquisition Management Information Retrieval database on March 31, 2016.

———, *Assessments for the October 2004 Review*, summary for F-22, PNO 265, October 2004b. Accessed via the Defense Acquisition Management Information Retrieval database on March 31, 2016.

———, *Assessments for the January 2005 Review*, summary for F/A-18E/F, PNO 549, January 2005a. Accessed via the Defense Acquisition Management Information Retrieval database on March 31, 2016.

———, *Assessments for the May 2005 Review*, summary for C-130J, PNO 220, May 2005b. Accessed via the Defense Acquisition Management Information Retrieval database on March 31, 2016.

———, *Assessments for the July 2005 Review*, summary for F-22, PNO 265, July 2005c. Accessed via the Defense Acquisition Management Information Retrieval database on March 31, 2016.

———, *Assessments for the November 2005 Review*, summary for C-130J, PNO 220, November 2005d. Accessed via the Defense Acquisition Management Information Retrieval database on March 31, 2016.

———, *Assessments for the February 2006 Review*, summary for C-17A, PNO 200, February 2006a. Accessed via the Defense Acquisition Management Information Retrieval database on March 31, 2016.

———, *Assessments for the May 2006 Review*, summary for C-17A, PNO 200, May 2006b. Accessed via the Defense Acquisition Management Information Retrieval database on March 31, 2016.

———, *Assessments for the August 2006 Review*, summary for C-130J, PNO 220, August 2006c. Accessed via the Defense Acquisition Management Information Retrieval database on March 31, 2016.

———, *Assessments for the October 2006 Review*, summary for F-22, PNO 265, October 2006d. Accessed via the Defense Acquisition Management Information Retrieval database on March 31, 2016.

———, *Assessments for the February 2013 Review*, summary for C-130J, PNO 220, February 2013a. Accessed via the Defense Acquisition Management Information Retrieval database, March 31, 2016.

———, *Assessments for the April 2013 Review*, summary for F/A-18E/F, PNO 549, April 2013b. Accessed via the Defense Acquisition Management Information Retrieval database on March 31, 2016.

———, *Assessments for the February 2015 Review*, summary for C-130J, PNO 220, February 2015. Accessed via the Defense Acquisition Management Information Retrieval database, March 31, 2016.

Defense Contract Management Agency, "Leveling the Playing Field, Fail-Safe RFP Proposal Instructions," undated briefing presented to RAND, Arlington, Va., March 20, 2015.

DoD—*See* U.S. Department of Defense.

Elmendorf, Douglas, informational letter regarding Navy Littoral Combat Ship procurement plans, Washington D.C., December 10, 2010, As of February 23, 2016:
https://www.cbo.gov/sites/default/files/111th-congress-2009-2010/reports/12-09_mccain_letter_final.pdf

F-16 System Program Office, *Validation of Multiyear Savings Associated with the Production of 720 F-16 Aircraft (FY86–FY89 Requirements)*, unpublished paper, September 1986.

"The F-22 Raptor: Program and Events," *Defense Industry Daily*, March 14, 2016.

Foote, Sheila, "Appropriators Back C-17 Multiyear Buy, with Conditions," *Defense Daily*, April 1, 1996a.

———, "Senate Passes Bill Authorizing C-17 Multiyear Buy," *Defense Daily*, May 24, 1996b.

GAO—*See* U.S. General Accounting Office (and U.S. Government Accountability Office).[142]

Grossman, Elaine M., "Air Force Finds Cracks in F-22 Fighter That May Prompt Tail Redesign," *Inside the Air Force*, August 10, 2001.

Harmon, Bruce R., Scot A. Arnold, James A. Myers, J. Richard Nelson, John R. Hiller, M. Michael Metcalf, Harold S. Balban, and Harley A. Cloud, "F-22A Multiyear Procurement Business Case Analysis," Institute for Defense Analyses, P-4116, undated (circa 2006).

Hebert, Adam, "Air Force Declines Lockheed Offer for Conditional Acceptance of C-130J," *Inside the Air Force*, January 8, 1999.

———, "Canopy Cracks Ground Raptor Fleet; Replacements Are on the Way," *Inside the Air Force*, May 26, 2000.

Heisler, Timrek, *C-130 Hercules: Background, Sustainment, Modernization, Issues for Congress*, Washington, D.C: Congressional Research Service, R43618, June 24, 2014.

Judson, Jen, "IG: Boeing Overcharged for CH-47F Parts and Inefficiently Used Funds," *Inside the Army*, July 22, 2013.

Kolcum, Edward H., "Lockheed Weighs Investment Risks of Developing New C-130 Version," *Aviation Week & Space Technology*, Vol. 131, No. 22, November 27, 1989.

"Lockheed Martin Snags $4 Billion Contract for C-130J Multiyear Deal," *Inside the Air Force*, March 21, 2003.

Lockheed Martin Aeronautics Company, "U.S. Government, Lockheed Martin Announce C-130J Super Hercules Multiyear II Contract," *PR Newswire*, December 31, 2015. As of January 14, 2016:
http://www.prnewswire.com/news-releases/us-government-lockheed-martin-announce-c-130j-super-hercules-multiyear-ii-contract-300198244.html

"Lockheed Officials Say C-130J Upgrades Are Effective and on Schedule," *Inside the Air Force*, June 11, 1999.

---

[142] This office changed its name in 2004.

Lorell, Mark A., John C. Graser, and Cynthia Cook, *Price-Based Acquisition: Issues and Challenges for Defense Department Procurement of Weapon Systems*, Santa Monica, Calif.: RAND Corporation, MG-337-AF, 2005. As of July 13, 2017: https://www.rand.org/pubs/monographs/MG337.html

Nichiporuk, Brian, *Alternative Futures and Army Force Planning*, Santa Monica, Calif.: RAND Corporation, MG-219-A, 2005. As of July 13, 2017: http://www.rand.org/pubs/monographs/MG219.html

Office of the Under Secretary of Defense for Acquisition, Technology, and Logistics, *Report of the Price-Based Acquisition Study Group*, November 15, 1999.

O'Rourke, Ronald, and Moshe Schwartz, *Multiyear Procurement (MYP) and BB Contracting in Defense Acquisition: Background and Issues for Congress*, Washington, D.C.: Congressional Research Service, R-41909, March 4, 2015.

"Pentagon Plans to Terminate C-130J Procurement in FY-06 Budget," *Inside the Air Force*, January 7, 2005.

"Senior USAF Officer: C-130J Termination Costs Could Exceed $2 Billion," *Inside the Air Force*, March 4, 2005.

Sherman, Jason, "Higher CAPE Estimate Amplifies Army Savings Claim in New CH-47F Contract," *Inside the Pentagon*, July 4, 2013.

Sobie, Brendan, "Air Force to Postpone High-Rate C-130J Acquisition Until at Least FY-03," *Inside the Air Force*, April 10, 1998.

Taylor, Dan, "Navy Awards $5.3 Billion Contract to Boeing for F/A-18, EA-18G Multiyear," *Inside the Navy*, October 4, 2010.

U.S. Air Force, *FY 2007 Budget Estimates, Aircraft Procurement, Air Force*, Vol. 1, February 2006.

———, *Air Force Signs Multiyear Contract for F-22*, August 8, 2007. As of March 31, 2016: http://www.af.mil/news/articledisplay/tabid/223/article/126070/air-force-signs-multiyear-contract-for-f-22.aspx

U.S. Department of Defense, *C-17A Globemaster III*, Selected Acquisitions Reports, executive summaries, assorted dates.

———, *Selected Acquisition Report: F/A-18E/F Super Hornet Aircraft (F/A-18E/F)*, December 2001a.

———, *Selected Acquisition Report: F-22 Raptor Advanced Tactical Fighter Aircraft (F-22)*, December 2001b.

———, *Selected Acquisition Report: CH-47F Improved Cargo Helicopter*, 2003.

————, *Selected Acquisition Report: CH-47F Improved Cargo Helicopter*, 2004.

————, *Selected Acquisition Report: C-130J Hercules Transport Aircraft (C-130J)*, December 2006a.

—————, *Selected Acquisition Report: F-22 Raptor Advanced Tactical Fighter Aircraft (F-22)*, December 2006b.

————, *Selected Acquisition Report: CH-47F Improved Cargo Helicopter*, December 2007.

————, *Selected Acquisitions Report: CH-47F Improved Cargo Helicopter (CH-47F)*, December 2009.

————, *Selected Acquisition Report: FP-EPA Efforts $958.8 Million*, December 2010.

————, *NDAA Sectional Analysis*, March 12, 2012.

————, *Selected Acquisition Report: CH-47F Improved Cargo Helicopter*, December 2014a.

————, *Selected Acquisition Report: MH-60R Multi-Mission Helicopter (MH-60R)*, December 2014b.

————, *Selected Acquisition Report: MH-60S Fleet Combat Support Helicopter (MH-60S)*, Washington, D.C., December 2014c.

————, *Selected Acquisition Report: C-130J SAR*, Washington, D.C., December 2015.

U.S. Department of Defense Office of the Inspector General, *Acquisition: Contracting for and Performance of the C-130J Aircraft*, Washington, D.C., D-2004-102, July 23, 2004.

—————, *Review of Defense Contract Management Agency Support of the C-130J Aircraft Program*, Report No. D-2009-074, June 12, 2009.

U.S. General Accounting Office, *An Assessment of the Air Force's F-16 Aircraft Multiyear Contract*, Washington, D.C., GAO/NSIAD-86-38, February 1986.

————, *C-17 Aircraft: Comments on Air Force Request for Approval of Multiyear Procurement Authority*, Washington, D.C., GAO/T-NSIAD-96-137, March 1996.

————, *Defense Acquisitions: Progress of the F/A-18E/F Engineering and Manufacturing Development Program*, Washington, D.C., GAO/NSIAD-99-127, June 1999.

————, *Defense Acquisitions: Recent F-22 Production Cost Estimates Exceeded Congressional Limitation*, Washington, D.C.: GAO/NSIAD-00-178, August 2000.

U.S. Government Accountability Office, *Defense Acquisitions: Assessments of Selected Weapon Programs*, Washington, D.C., GAO-07-406SP, March 2007.

————, *DoD's Practices and Processes for Multiyear Procurement Should Be Improved*, Washington, D.C., GAO-08-298, February 2008.

U.S. House of Representatives, *Approving Purchases of Littoral Combat Ships*, Vol. 156, No. 166, December 15, 2010, pp. H8359–H8362, As of February 23, 2016: https://www.congress.gov/crec/2010/12/15/CREC-2010-12-15-pt1-PgH8359-5.pdf

U.S. Senate, *National Defense Authorization Act for Fiscal Year 1998 Report*, Report 105-29, June 17, 1997. As of February 23, 2016: https://www.congress.gov/105/crpt/srpt29/CRPT-105srpt29.pdf

———, *National Defense Authorization Act for Fiscal Year 2001 Report*, 106-292, May 12, 2000. As of May 31, 2016: https://www.congress.gov/106/crpt/srpt292/CRPT-106srpt292.pdf

———, *National Defense Authorization Act for Fiscal Year 2003 Report*, Report 107-151, May 15, 2002, As of February 23, 2016: https://www.congress.gov/107/crpt/srpt151/CRPT-107srpt151.pdf

Younossi, Obaid, Mark V. Arena, Kevin Brancato, John C. Graser, Benjamin W. Goldsmith, Mark A. Lorell, Fred Timson, and Jerry M. Sollinger, *F-22A Multiyear Procurement Program: An Assessment of Cost Savings*, Santa Monica, Calif.: RAND Corporation, MG-664-OSD, 2007. As of February 23, 2016: http://www.rand.org/pubs/monographs/MG664.html

Younossi, Obaid, David E. Stem, Mark A. Lorell, and Frances M. Lussier, *Lessons Learned from the F-22 and F/A-18E/F Development Programs*, Santa Monica, Calif.: RAND Corporation, MG-276-AF, 2005. As of February 23, 2016: http://www.rand.org/pubs/monographs/MG276.html